FEMA 403 / *May 2002*

World Trade Center Building Performance Study:

Data Collection, Preliminary Observations, and Recommendations

Federal Emergency Management Agency

Federal Insurance and Mitigation Administration, Washington, DC

FEMA Region II, New York, New York

Report Editor:
Therese McAllister

Chapter Leaders and Authors:

Executive Summary Gene Corley
　　Ronald Hamburger
　　Therese McAllister

Chapter 1 Therese McAllister
　　Jonathan Barnett
　　John Gross
　　Ronald Hamburger
　　Jon Magnusson

Chapter 2 Ronald Hamburger
　　William Baker
　　Jonathan Barnett
　　Christopher Marrion
　　James Milke
　　Harold "Bud" Nelson

Chapter 3 William Baker

Chapter 4 Jonathan Barnett
　　Richard Gewain
　　Ramon Gilsanz
　　Harold "Bud" Nelson

Chapter 5 Ramon Gilsanz
　　Edward M. DePaola
　　Christopher Marrion
　　Harold "Bud" Nelson

Chapter 6 Robert Smilowitz
　　Adam Hapij
　　Jeffrey Smilow

Chapter 7 Therese McAllister
　　David Biggs
　　Edward M. DePaola
　　Dan Eschenasy
　　Ramon Gilsanz

Appendix A James Milke
　　Venkatesh Kodur
　　Christopher Marrion

Appendix B John Fisher
　　Nestor Iwankiw

Appendix C Jonathan Barnett
　　Ronald R. Biederman
　　R. D. Sisson, Jr.

Appendix D Ramon Gilsanz
　　Audrey Massa

Appendix F Edward M. DePaola

This report was produced by Greenhorne & O'Mara, Inc. (G&O) under Contract EMW-99-CO-0267 to the Federal Insurance and Mitigation Administration of FEMA

G&O Project Manager:	Eric Letvin
FEMA Project Officer:	Paul Tertell
ASCE Project Manager:	James Rossberg
G&O Technical Editors:	Bob Pendley
	Deb Daly
G&O Graphic Designers:	Wanda Rizer
	Julie Liptak

Team Leader:
Gene Corley

Team Members:

William Baker, Partner,
Skidmore Owings & Merrill LLP

Jonathan Barnett, Professor of Fire Protection Engineering,
Worcester Polytechnic Institute

David Biggs, Principal, Ryan-Biggs Associates

Gene Corley, Structural Engineer, Senior Vice President,
Construction Technology Laboratories

Bill Coulbourne, Principal Structural Engineer,
URS Corporation

Edward M. DePaola, Principal,
Severud Associates Consulting Engineers, PC

Robert Duval, Senior Fire Investigator,
National Fire Protection Association

Dan Eschenasy, Chief Structural Engineer,
City of New York Dept. of Design & Construction

John Fisher, Professor of Civil Engineering,
Lehigh University

Richard Gewain, Senior Engineer,
Hughes Associates, Inc.

Ramon Gilsanz, Partner, Gilsanz Murray Steficek

John Gross, Structural Systems and Design Group Leader,
National Institute of Standards & Technology

Ronald Hamburger, Chief Structural Engineer,
ABS Consulting

Nestor Iwankiw, Vice President, Engineering and Research,
American Institute of Steel Construction

Venkatesh Kodur, Research Officer,
National Research Council of Canada

Eric Letvin, Director, Hazards Mitigation Division,
Greenhorne & O'Mara, Inc.

Jon Magnusson, Partner,
Skilling Ward Magnusson Barkshire, Inc.

Christopher Marrion, Associate, Arup Fire

Therese McAllister, Senior Structural Engineer,
Greenhorne & O'Mara, Inc.

James Milke, Professor of Fire Protection,
University of Maryland

Harold E. "Bud" Nelson, Senior Research Engineer,
Hughes Associates, Inc.

James Rossberg, Director,
Structural Engineering Institute

Saw-Teen See, Managing Partner,
Leslie E. Robertson Associates, RLLP

Robert Smilowitz, Principal, Weidlinger Associates

Bruce Swiren, Senior EMP Specialist, FEMA Region II

Paul Tertell, Civil Engineer, FEMA

Title page photograph courtesy of Val Junker, Mobius Communications, Inc., Princeton, NJ.

Table of Contents

Executive Summary

1 Introduction
1.1 Purpose and Scope of Study 1-1
1.2 WTC Site 1-2
1.3 Timeline and Event Summary 1-4
1.4 Response of the Engineering Community 1-8
    1.4.1 Local Authorities 1-8
    1.4.2 SEAoNY Participation 1-10
1.5 Overview of Building Codes and Fire Standards 1-15
    1.5.1 Building Codes 1-15
    1.5.2 Unusual Building Loads 1-16
    1.5.3 Overview of Fire-Structure Interaction 1-17
        1.5.3.1 ASTM E119 Standard Fire Test 1-18
        1.5.3.2 Performance in Actual Building Fires 1-18
1.6 Report Organization 1-20
1.7 References 1-21

2 WTC 1 and WTC 2
2.1 Building Descriptions 2-1
    2.1.1 General 2-1
    2.1.2 Structural Description 2-1
    2.1.3 Fire Protection 2-11
        2.1.3.1 Passive Protection 2-12
        2.1.3.2 Suppression 2-12

FEDERAL EMERGENCY MANAGEMENT AGENCY

Table of Contents

    2.1.3.3 Smoke Management 2-13
    2.1.3.4 Fire Department Features 2-13
  2.1.4 Emergency Egress 2-13
  2.1.5 Emergency Power 2-14
  2.1.6 Management Procedures 2-14

2.2 Building Response 2-15
  2.2.1 WTC 1 2-15
    2.2.1.1 Initial Damage From Aircraft Impact 2-15
    2.2.1.2 Fire Development 2-21
    2.2.1.3 Evacuation 2-24
    2.2.1.4 Structural Response to Fire Loading 2-24
    2.2.1.5 Progression of Collapse 2-27
  2.2.2 WTC 2 2-27
    2.2.2.1 Initial Damage From Aircraft Impact 2-27
    2.2.2.2 Preliminary Structural Analysis 2-32
    2.2.2.3 Fire Development 2-34
    2.2.2.4 Evacuation 2-35
    2.2.2.5 Initiation of Collapse 2-35
    2.2.2.6 Progression of Collapse 2-35
  2.2.3 Substructure 2-35

2.3 Observations and Findings 2-36
2.4 Recommendations 2-39
2.5 References 2-40

3 WTC 3

3.1 Design and Construction Features 3-1
  3.1.1 Project Overview 3-1
  3.1.2 Building Description 3-1
  3.1.3 Structural Description 3-2

3.2 1993 Attack 3-5
3.3 2001 Attacks 3-5
  3.3.1 Fire and Evacuation 3-5
  3.3.2 Building Response 3-6

3.4 Observations 3-8
3.5 Recommendations 3-8
3.5 References 3-8

Table of Contents

4 WTC 4, 5, and 6

4.1	Design and Construction Features	4-1
	4.1.1 Structural Design Features	4-1
	4.1.2 Fire Protection Features	4-2
4.2	Building Loads and Performance	4-4
	4.2.1 Impact Damage to WTC 5	4-7
	4.2.2 Fire Damage	4-9
4.3	Analysis of Building Performance	4-10
	4.3.1 Steel and Frame Behavior	4-10
	4.3.2 WTC 5 – Local Collapse Mechanisms	4-12
4.4	Observations and Findings	4-16
4.5	Recommendations	4-21
4.6	References	4-21

5 WTC 7

5.1	Introduction	5-1
5.2	Structural Description	5-3
	5.2.1 Foundations	5-3
	5.2.2 Structural Framing	5-4
	5.2.3 Transfer Trusses and Girders	5-4
	5.2.4 Connections	5-8
5.3	Fire Protection Systems	5-10
	5.3.1 Egress Systems	5-10
	5.3.2 Detection and Alarm	5-10
	5.3.3 Compartmentalization	5-11
	5.3.4 Suppression Systems	5-12
	5.3.5 Power	5-13
5.4	Building Loads	5-13
5.5	Timeline of Events Affecting WTC 7 on September 11, 2001	5-16
	5.5.1 Collapse of WTC 2	5-16
	5.5.2 Collapse of WTC 1	5-16
	5.5.3 Fires at WTC 7	5-20
	5.5.4 Sequence of WTC 7 Collapse	5-23

Table of Contents

    5.6   Potential Collapse Mechanism 5-24

            5.6.1   Probable Collapse Initiation Events 5-24

            5.6.2   Probable Collapse Sequence 5-30

    5.7   Observations and Findings 5-31

    5.8   Recommendations 5-32

    5.9   References 5-32

6   Bankers Trust Building

    6.1   Introduction 6-1

    6.2   Building Description 6-1

    6.3   Structural Damage Description 6-4

    6.4   Architectural Damage Description 6-6

    6.5   Fireproofing 6-10

    6.6   Overall Assessment 6-10

    6.7   Analysis 6-11

            6.7.1   Key Assumptions 6-12

            6.7.2   Model Refinement 6-12

            6.7.3   Simulation of Nonlinear Behavior 6-13

            6.7.4   Connection Details 6-13

            6.7.5   Connection Behavior 6-14

    6.8   Observations and Findings 6-15

    6.9   Recommendations 6-16

    6.10 References 6-16

7   Peripheral Buildings

    7.1   Introduction 7-1

    7.2   World Financial Center 7-4

            7.2.1   The Winter Garden 7-4

            7.2.2   WFC 3, American Express Building 7-5

    7.3   Verizon Building 7-7

    7.4   30 West Broadway 7-13

    7.5   130 Cedar Street 7-14

    7.6   90 West Street 7-15

Table of Contents

7.7	45 Park Place	7-17
7.8	One Liberty Plaza	7-17
7.9	Observations and Findings	7-19
7.10	Recommendations	7-19
7.11	References	7-19

8   Observations, Findings, and Recommendations

8.1	Summary of Report Observations, Findings, and Recommendations	8-1
8.2	Chapter Observations, Findings, and Recommendations	8-2
	8.2.1 Chapter 1: Building Codes and Fire Standards	8-2
	8.2.2 Chapter 2: WTC 1 and WTC 2	8-2
	8.2.2.1 Observations and Findings	8-2
	8.2.2.2 Recommendations	8-5
	8.2.3 Chapter 3: WTC 3	8-6
	8.2.3.1 Observations	8-6
	8.2.3.2 Recommendations	8-6
	8.2.4 Chapter 4: WTC 4, 5, and 6	8-6
	8.2.4.1 Observations and Findings	8-6
	8.2.4.2 Recommendations	8-7
	8.2.5 Chapter 5: WTC 7	8-7
	8.2.5.1 Observations and Findings	8-7
	8.2.5.2 Recommendations	8-8
	8.2.6 Chapter 6: Bankers Trust	8-8
	8.2.6.1 Observations and Findings	8-8
	8.2.6.2 Recommendations	8-9
	8.2.7 Chapter 7: Peripheral Buildings	8-10
	8.2.7.1 Observations and Findings	8-10
	8.2.7.2 Recommendations	8-10
	8.2.8 Appendix C: Limited Metallurgical Examination	8-10
	8.2.8.1 Observations and Findings	8-10
	8.2.8.2 Recommendations	8-11
8.3	Building Performance Study Recommendations for Future Study	8-11
	8.3.1 National Response	8-11
	8.3.2 Interaction of Structural Elements and Fire	8-11
	8.3.3 Interaction of Professions in Design	8-12
	8.3.4 Fire Protection Engineering Discipline	8-12
	8.3.5 Building Evacuation	8-12

Table of Contents

    8.3.6   Emergency Personnel 8-12

    8.3.7   Education of Stakeholders 8-13

    8.3.8   Study Process 8-13

    8.3.9   Archival Information 8-13

    8.3.10   SEAoNY Structural Engineering Emergency Response Plan 8-13

A   Overview of Fire Protection in Buildings

A.1 Introduction A-1

A.2 Fire Behavior A-1

    A.2.1   Burning Behavior of Materials A-1

    A.2.2   Stages of Fire Development A-4

    A.2.3   Behavior of Fully Developed Fires A-4

A.3 Structural Response to Fire A-5

    A.3.1   Effect of Fire on Steel A-5

        A.3.1.1   Introduction A-5

        A.3.1.2   Evaluating Fire Resistance A-6

        A.3.1.3   Response of High-rise, Steel-frame Buildings in Previous Fire Incidents A-9

        A.3.1.4   Properties of Steel A-10

        A.3.1.5   Fire Protection Techniques for Steel A-14

        A.3.1.6   Temperature Rise in Steel A-14

        A.3.1.7   Factors Affecting Performance of Steel Structures in Fire A-17

    A.3.2   Effect of Fire on Concrete A-19

        A.3.2.1   General A-19

        A.3.2.2   Properties of Lightweight Concrete A-19

    A.3.3   Fire and Structural Modeling A-22

A.4 Life Safety A-23

    A.4.1   Evacuation Process A-24

    A.4.2   Analysis A-24

A.5 References A-26

B   Structural Steel and Steel Connections

B.1 Structural Steel B-1

B.2 Mechanical Properties B-2

B.3 WTC 1 and WTC 2 Connection Capacity B-4

    B.3.1   Background B-4

Table of Contents

	B.3.2	Observations B-5
	B.3.3	Connectors B-5
B.4	Examples of WTC 1 and WTC 2 Connection Capacity B-7	
	B.4.1	Bolted Column End Plates B-7
	B.4.2	Bolted Spandrel Connections B-8
	B.4.3	Floor Truss Seated End Connections at Spandrel Beam and Core B-9
	B.4.4	WTC 5 Column-tree Shear Connections B-12
B.5	References B-14	

C   Limited Metallurgical Examination

C.1   Introduction C-1

C.2   Sample 1 (From WTC 7) C-1

C.3   Summary for Sample 1 .... C-5

C.4   Sample 2 (From WTC 1 or WTC 2) C-5

C.5   Summary for Sample 2 .... C-13

C.6   Suggestions for Future Research C-13

D   WTC Steel Data Collection

D.1   Introduction D-1

D.2   Project Background D-1

D.3   Methods D-2

    D.3.1   Identifying and Saving Pieces D-2

    D.3.2   Documenting Pieces D-5

    D.3.3   Getting Coupons D-8

D.4   Data Collected D-10

D.5   Conclusions and Future Work D-13

D.6   References D-13

E   Aircraft Information

General Specifications E-1

F   Structural Engineers Emergency Response Plan

G   Acknowledgments

FEDERAL EMERGENCY MANAGEMENT AGENCY

Table of Contents

H	Acronyms and Abbreviations

I	Metric Conversions

Tables

Table 1.1	Timeline of Major Events	1-10
Table 2.1	Estimated Openings in Exterior Walls of WTC 1	2-23
Table 5.1	WTC 7 Tenants	5-2
Table 5.2	WTC 7 Fuel Distribution Systems	5-14
Table 7.1	DoB/SEAoNY Cooperative Building Damage Assessment – November 7, 2001	7-3
Table A.1	Peak Heat Release Rates of Various Materials (NFPA 92B and NFPA 72)	A-3
Table A.2	Fire Duration in Previous Fire Incidents in Steel-frame Buildings	A-10
Table A.3	Critical Temperatures for Various Types of Steel	A-15
Table A.4	Test Methods for Spray-applied Fireproofing Materials	A-18

Figures

Chapter 1

Figure 1-1	WTC site map.	1-3
Figure 1-2	Approximate flight paths of aircraft.	1-5
Figure 1-3	WTC impact locations and resulting fireballs.	1-5
Figure 1-4	Areas of aircraft debris impact.	1-6
Figure 1-5	Fireball erupts on the north face of WTC 2 as United Airlines Flight 175 strikes the building.	1-7
Figure 1-6	View of the north and east faces showing fire and impact damage to both towers.	1-7
Figure 1-7	Schematic depiction of areas of collapse debris impact, based on aerial photographs and documented damage.	1-9
Figure 1-8	Seismic recordings on east-west component at Palisades, NY, for events at WTC on September 11, 2001, distance 34 km.	1-11
Figure 1-9A	Satellite photograph of the WTC site taken before the attacks.	1-12
Figure 1-9B	Satellite photograph of the WTC site taken after the attacks.	1-13
Figure 1-10	Comparison of high-rise building and aircraft sizes.	1-19

Table of Contents

Chapter 2

Figure 2-1	Representative floor plan (based on floor plan for 94th and 95th floors of WTC 1).	2-2
Figure 2-2	Representative structural framing plan, upper floors.	2-4
Figure 2-3	Partial elevation of exterior bearing-wall frame showing exterior wall module construction.	2-6
Figure 2-4	Base of exterior wall frame.	2-7
Figure 2-5	Structural tube frame behavior.	2-7
Figure 2-6	Floor truss member with details of end connections.	2-8
Figure 2-7	Erection of prefabricated components, forming exterior wall and floor deck units.	2-9
Figure 2-8	Erection of floor framing during original construction.	2-9
Figure 2-9	Cross-section through main double trusses, showing transverse truss.	2-9
Figure 2-10	Outrigger truss system at tower roof.	2-10
Figure 2-11	Location of subterranean structure.	2-11
Figure 2-12	Floor plan of 94th and 95th floors of WTC 1 showing egress stairways.	2-14
Figure 2-13	Zone of aircraft impact on the north face of WTC 1.	2-16
Figure 2-14	Approximate zone of impact of aircraft on the north face of WTC 1.	2-17
Figure 2-15	Impact damage to the north face of WTC 1.	2-18
Figure 2-16	Impact damage to exterior columns on the north face of WTC 1.	2-18
Figure 2-17	Approximate debris location on the 91st floor of WTC 1.	2-19
Figure 2-18	Landing gear found at the corner of West and Rector Streets	2-19
Figure 2-19	Redistribution of load after aircraft impact.	2-20
Figure 2-20	Expansion of floor slabs and framing results in outward deflection of columns and potential overload.	2-25
Figure 2-21	Buckling of columns initiated by failure of floor joist connections.	2-26
Figure 2-22	Catenary action of floor framing on several floors initiates column buckling failures.	2-26
Figure 2-23	Aerial photograph of the WTC site after September 11 attack showing adjacent buildings damaged by debris from the collapse of WTC 1.	2-28
Figure 2-24	Southeast corner of WTC 2 shortly after aircraft impact.	2-28
Figure 2-25	Approximate zone of impact of aircraft on the south face of WTC 2.	2-29
Figure 2-26	Impact damage to the south and east faces of WTC 2.	2-30
Figure 2-27	Impact damage to exterior columns on the south face of WTC 2.	2-30

Table of Contents

Figure 2-28	Conflagration and debris exiting the north wall of WTC 2, behind WTC 1.	2-31
Figure 2-29	A portion of the fuselage of United Airlines Flight 175 on the roof of WTC 5.	2-32
Figure 2-30	North face of WTC 2 opposite the zone of impact on the south face, behind WTC 1.	2-33
Figure 2-31	Plot of column utilization ratio at the 80th floor of WTC 2, viewed looking outward.	2-34
Figure 2-32	The top portion of WTC 2 falls to the east, then south, as viewed from the northeast.	2-36

Chapter 3

Figure 3-1	Developed north and west elevations.	3-2
Figure 3-2	Developed south and east elevations.	3-2
Figure 3-3	Typical hotel floor plan.	3-3
Figure 3-4	Typical hotel floor framing plan.	3-4
Figure 3-5	Typical transverse bracing elevation.	3-5
Figure 3-6	Exterior columns from the collapse of WTC 2 falling on the southern part of WTC 3.	3-6
Figure 3-7	Partial collapse of WTC 3 after collapse of WTC 2.	3-7
Figure 3-8	Remains of WTC 3 after collapse of WTC 1 and WTC 2.	3-8

Chapter 4

Figure 4-1	Typical floor plan for WTC 5.	4-2
Figure 4-2	Typical column-tree system (not to scale).	4-3
Figure 4-3	Typical interior bay framing in WTC 5.	4-3
Figure 4-4	Stairway enclosure core locations in WTC 5.	4-4
Figure 4-5	Damage to WTC 4.	4-5
Figure 4-6	Damage to WTC 5.	4-5
Figure 4-7	Approximate locations of damaged floor areas in WTC 5.	4-6
Figure 4-8	Damage to WTC 6.	4-8
Figure 4-9	Impact damage to WTC 6.	4-8
Figure 4-10	Impact damage to the exterior façade of WTC 6.	4-9
Figure 4-11	WTC 5 façade damage.	4-10
Figure 4-12	Impact damage to WTC 5.	4-11

Table of Contents

Figure 4-13	WTC 5 on fire.	4-13
Figure 4-14	Deformed beams in WTC 5.	4-13
Figure 4-15	Unburned bookstore in WTC 5.	4-14
Figure 4-16	Looking through the door into the undamaged stair tower in WTC 5.	4-14
Figure 4-17	Buckled beam flange and column on the 8th floor of WTC 5 that was weakened by fire.	4-15
Figure 4-18	Internal collapsed area in WTC 5.	4-16
Figure 4-19	Internal collapsed area in WTC 5.	4-17
Figure 4-20	Internal collapsed area in WTC 5.	4-18
Figure 4-21	Internal collapsed area in WTC 5 with closeup of connection failure at column tree.	4-18
Figure 4-22	Connection samples.	4-19

Chapter 5

Figure 5-1	Foundation plan – WTC 7.	5-3
Figure 5-2	Plan view of typical floor framing for the 8th through 45th floors.	5-4
Figure 5-3	Elevations of building and core area.	5-5
Figure 5-4	Fifth floor diaphragm plan showing T-sections embedded in 14-inch slab.	5-6
Figure 5-5	3-D diagram showing relations of trusses and transfer girders.	5-6
Figure 5-6	Seventh floor plan showing locations of transfer trusses and girders.	5-7
Figure 5-7	Truss 1 detail.	5-8
Figure 5-8	Truss 2 detail.	5-9
Figure 5-9	Truss 3 detail.	5-10
Figure 5-10	Cantilever transfer girder detail.	5-11
Figure 5-11	Compartmentalization provided by concrete floor slabs.	5-12
Figure 5-12	Sequence of debris generated by collapses of WTC 2, 1, and 7.	5-17
Figure 5-13	Pedestrian bridge (bottom center) still standing after WTC 2 has collapsed, sending substantial dust and debris onto the street, but before WTC 1 (top center) has collapsed.	5-18
Figure 5-14	View from the north of the WTC 1 collapse and the spread of debris around WTC 7.	5-18
Figure 5-15	Debris from the collapse of WTC 1 located between WTC 7 (left) and the Verizon building (right).	5-19

Table of Contents

Figure 5-16   Damage to the southeast corner of WTC 7 (see box), looking from West Street. 5-19

Figure 5-17   Building damage to the southwest corner and smoke plume from
               south face of WTC 7, looking from the World Financial Plaza. 5-20

Figure 5-18   WTC 7, with a large volume of dark smoke rising from it, just visible
               behind WFC 1 (left). 5-21

Figure 5-19   Fires on the 11th and 12th floors of the east face of WTC 7. 5-22

Figure 5-20   View from the north of WTC 7 with both mechanical penthouses intact. 5-24

Figure 5-21   East mechanical penthouse collapsed. 5-25

Figure 5-22   East and now west mechanical penthouses gone. 5-25

Figure 5-23   View from the north of the "kink" or fault developing in WTC 7. 5-26

Figure 5-24   Areas of potential transfer truss failure. 5-27

Figure 5-25   Debris cloud from collapse of WTC 7. 5-27

Figure 5-26   Debris generated after collapse of WTC 7. 5-28

Chapter 6

Figure 6-1   North face of Bankers Trust building with impact damage
               between floors 8 and 23. 6-1

Figure 6-2   Closeup of area of partial collapse. 6-2

Figure 6-3   Floor plan above the 2nd level (ground floor extension not shown). 6-3

Figure 6-4   Area of initial impact of debris at the 23rd floor. 6-4

Figure 6-5   Approximate zones of damage – 19th through 22nd floors, 16th through
               18th floors, 11th through 15th floors, and 9th through 10th floors. 6-5

Figure 6-6   Moment-connected beams to columns. 6-7

Figure 6-7   Column with the remains of two moment connections. 6-7

Figure 6-8   Failed shear connection of beam web to column web. 6-8

Figure 6-9   Suspended column D-8 at the 15th floor. 6-8

Figure 6-10   Area of collapsed floor slab in bays between C-8, E-8, C-7, and
               E-7, from 15th floor. 6-9

Figure 6-11   Bankers Trust lobby (note debris has been swept into piles). 6-9

Figure 6-12   Office at north side of the 8th floor. 6-10

Figure 6-13   3-D ANSYS model of flange and shear plate moment connection. 6-14

Figure 6-14   3-D ANSYS model of flange and seat moment connection. 6-14

Table of Contents

Chapter 7

Figure 7-1	New York City DDC/DoB Cooperative Building Damage Assessment Map of November 7, 2001 (based on SEAoNY inspections of September and October 2001).	7-2
Figure 7-2	Southeast corner of WFC 3.	7-5
Figure 7-3	View of Winter Garden damage from West Street, with WTC 1 debris in front of WFC 2.	7-6
Figure 7-4	View of Winter Garden damage from West Street, with WTC 1 debris leaning against WFC 3.	7-6
Figure 7-5	Interior damage at floor 20 of WFC 3.	7-7
Figure 7-6	Verizon building – damage to east elevation (Washington Street).	7-8
Figure 7-7	Verizon building – damage to east elevation (Washington Street) due to WTC 7 framing leaning against the building.	7-9
Figure 7-8	Verizon building – damage to east elevation (Washington Street).	7-9
Figure 7-9	Verizon building – column damage on east elevation (Washington Street).	7-10
Figure 7-10	Verizon building – damage to south elevation (Vesey Street).	7-11
Figure 7-11	Verizon building – localized damage to south elevation (Vesey Street).	7-11
Figure 7-12	Verizon building – detail of damage to south elevation (Vesey Street).	7-12
Figure 7-13	30 West Broadway – south façade, 6th floor to roof, looking northeast.	7-13
Figure 7-14	130 Cedar Street and 90 West Street.	7-14
Figure 7-15	Interior of 90 West Street showing typical construction features.	7-16
Figure 7-16	Buckling damage at top of column on floor 8 of 90 West Street.	7-17
Figure 7-17	Buckling damage at top of column on floor 23 of 90 West Street.	7-18
Figure 7-18	One Liberty Plaza – south elevation, lower floors.	7-19
Figure 7-19	One Liberty Plaza – south elevation, upper floors.	7-20

Appendix A

Figure A-1	Heat release rate for office module.	A-2
Figure A-2	Fire growth rates (from SFPE Handbook of Fire Protection Engineering).	A-3
Figure A-3	Comparison of exposure temperatures in standard tests.	A-6
Figure A-4	Thermal properties of steel at elevated temperatures.	A-11
Figure A-5	Stress-strain curves for structural steel (ASTM A36) at a range of temperatures.	A-11
Figure A-6	Strength of steel at elevated temperatures.	A-12

Table of Contents

Figure A-7   Modulus of elasticity at elevated temperatures for structural steels and steel reinforcing bars. A-13

Figure A-8   Reduction of the yield strength of cold-formed light-gauge steel at elevated temperatures. A-13

Figure A-9   Steel temperature rise due to fire exposure for unprotected steel column. A-16

Figure A-10 Steel temperature rise due to fire exposure for steel column protected with 1 inch of spray-applied fireproofing. A-16

Figure A-11 The effect of temperature on the modulus of elasticity strength of different types of concretes. A-20

Figure A-12 Reduction of the compressive strength of two lightweight concretes (one with natural sand) at elevated temperatures. A-21

Figure A-13 Usual ranges of variation for the volume-specific heat of normal-weight and lightweight concretes. A-21

Figure A-14 Specific flow rate as a function of density (SPFE Handbook of Fire Protection Engineering). A-25

Figure A-15 Estimated evacuation times for high-rise buildings. A-26

Appendix B

Figure B-1   Exterior column end plates. B-1

Figure B-2   Tensile stress-strain curves for three ASTM-designation steels. B-3

Figure B-3   Expanded yield portion of the tensile stress-strain curves. B-3

Figure B-4   Effect of high strain rate on shape of stress-strain diagram. B-4

Figure B-5   Column tree showing bolt bearing shear failures of spandrel connection. B-6

Figure B-6   Shear fracture failure of fillet welds connecting a W-shape column to a box core column. B-7

Figure B-7   Bent and fractured bolts at a column four-bolt connection. B-8

Figure B-8   Typical truss top chord connections to column/spandrel beam and to the core beam. B-10

Figure B-9   (A) Visco-elastic damper angles bolted to angle welded to spandrel plate and (B) failed bearing seat connection. B-11

Figure B-10 (A) Bracket plate welded to the column/spandrel plate and (B) horizontal plate brace with shear connectors welded to the failed bracket. B-11

Figure B-11 Shear failure of floor joist connections from column/spandrel plate. B-12

Appendix C

Figure C-1	Eroded A36 wide-flange beam.	C-1
Figure C-2	Closeup view of eroded wide-flange beam section.	C-2
Figure C-3	Mounted and polished severely thinned section removed from the wide-flange beam shown in Figure C-1.	C-2
Figure C-4	Optical microstructure near the steel surface.	C-3
Figure C-5	Another hot corrosion region near the steel surface (etched with 4 percent nital).	C-3
Figure C-6	Microstructure of A36 steel.	C-4
Figure C-7	Deep penetration of liquid into the steel.	C-4
Figure C-8	Qualitative chemical analysis.	C-5
Figure C-9	Qualitative chemical analysis.	C-6
Figure C-10	Grain boundary corrosion attack.	C-6
Figure C-11	Microstructure of a typical region showing the surface and grain boundary corrosion attack of Sample 2.	C-7
Figure C-12	Higher magnification of the region shown in Figure C-10.	C-7
Figure C-13	Regions where chemical analysis was performed.	C-8
Figure C-14	Gradient of sulfides into the steel from the oxide-metal interface.	C-12

Appendix D

Figure D-1	Mixed, unsorted steel upon delivery to salvage yard.	D-2
Figure D-2	Torch cutting of very large pieces into more manageable pieces of a few tons each.	D-3
Figure D-3	Pile of unsorted mixed steel (background), with sorted, large steel pieces (center) being lifted and cut into smaller pieces (left).	D-3
Figure D-4	Engineer climbing in unprocessed steel pile to inspect and mark promising pieces	D-4
Figure D-5	Stenciled markings on WTC 2 perimeter column from floors 68-71.	D-5
Figure D-6	Steel pieces marked "SAVE."	D-6
Figure D-7	Engineers measuring and recording steel piece dimensions.	D-6
Figure D-8	Engineer measuring spandrel plate thickness (t_s).	D-7
Figure D-9	Measurement of 1/4 inch for web thickness (t_w).	D-7
Figure D-10	Measured dimensions of the steel pieces.	D-8
Figure D-11	Burnt steel piece marked for cutting of coupon.	D-9

Table of Contents

Figure D-12　Coupon cut from WTC 5 showing web tear-out at bolts. D-9

Figure D-13　WTC 1 or WTC 2 core column (C-74). D-10

Figure D-14　WTC 7 W14 column tree with beams attached to two floors. D-11

Figure D-15　Built-up member with failure along stitch welding. D-11

Figure D-16　Engineer inspecting fire damage of perimeter column tree from
　　　　　　　WTC 1 or WTC 2. .... D-12

Figure D-17　Seat-connected in fire-damaged W14 column from WTC 7. .... D-12

Figure D-18　WTC 1 or WTC 2 floor-truss section with seat connection
　　　　　　　fractured along welds. D-13

Executive Summary

Following the September 11, 2001, attacks on New York City's World Trade Center (WTC), the Federal Emergency Management Agency (FEMA) and the Structural Engineering Institute of the American Society of Civil Engineers (SEI/ASCE), in association with New York City and several other Federal agencies and professional organizations, deployed a team of civil, structural, and fire protection engineers to study the performance of buildings at the WTC site.

The events following the attacks in New York City were among the worst building disasters in history and resulted in the largest loss of life from any single building collapse in the United States. Of the 58,000 people estimated to be at the WTC Complex, 2,830 lost their lives that day, including 403 emergency responders. Two commercial airliners were hijacked, and each was flown into one of the two 110-story towers. The structural damage sustained by each tower from the impact, combined with the ensuing fires, resulted in the total collapse of each building. As the towers collapsed, massive debris clouds consisting of crushed and broken building components fell onto and blew into surrounding structures, causing extensive collateral damage and, in some cases, igniting fires and causing additional collapses. In total, 10 major buildings experienced partial or total collapse and approximately 30 million square feet of commercial office space was removed from service, of which 12 million belonged to the WTC Complex.

The purpose of this study was to examine the damage caused by these events, collect data, develop an understanding of the response of each affected building, identify the causes of observed behavior, and identify studies that should be performed. The immediate effects of the aircraft impacts on each tower, the spread of fires following the crashes, the fire-induced reduction of structural strength, and the mechanism that led to the collapse of each tower were studied. Additionally, the performance of buildings in the immediate vicinity of the towers was studied to determine the effects of damage from falling debris and fires. Recommendations are presented for more detailed engineering studies, to complete the assessments and produce improved guidance and tools for building design and performance evaluation.

As each tower was struck, extensive structural damage, including localized collapse, occurred at the several floor levels directly impacted by the aircraft. Despite this massive localized damage, each structure remained standing. However, as each aircraft impacted a building, jet fuel on board ignited. Part of this fuel immediately burned off in the large fireballs that erupted at the impact floors. Remaining fuel flowed across the floors and down elevator and utility shafts, igniting intense fires throughout upper portions of the buildings. As these fires spread, they further weakened the steel-framed structures, eventually leading to total collapse.

The collapse of the towers astonished most observers, including knowledgeable structural engineers, and, in the immediate aftermath, a wide range of explanations were offered in an attempt to help the public understand these tragic events. However, the collapse of these symbolic buildings entailed a complex series

Executive Summary

of events that were not identical for each tower. To determine the sequence of events, likely root causes, and methods or technologies that may improve or mitigate the building performance observed, FEMA and ASCE formed a Building Performance Study (BPS) Team consisting of specialists in tall building design, steel and connection technology, fire and blast engineering, and structural investigation and analysis.

The Team conducted field observations at the WTC site and steel salvage yards, removed and tested samples of the collapsed structures, viewed hundreds of hours of video and thousands of still photographs, conducted interviews with witnesses and persons involved in the design, construction, and maintenance of each of the affected buildings, reviewed construction documents, and conducted preliminary analyses of the damage to the WTC towers.

With the information and time available, the sequence of events leading to the collapse of each tower could not be definitively determined. However, the following observations and findings were made:

The structural damage sustained by each of the two buildings as a result of the terrorist attacks was massive. The fact that the structures were able to sustain this level of damage and remain standing for an extended period of time is remarkable and is the reason that most building occupants were able to evacuate safely. Events of this type, resulting in such substantial damage, are generally not considered in building design, and the ability of these structures to successfully withstand such damage is noteworthy.

Preliminary analyses of the damaged structures, together with the fact the structures remained standing for an extended period of time, suggest that, absent other severe loading events such as a windstorm or earthquake, the buildings could have remained standing in their damaged states until subjected to some significant additional load. However, the structures were subjected to a second, simultaneous severe loading event in the form of the fires caused by the aircraft impacts.

The large quantity of jet fuel carried by each aircraft ignited upon impact into each building. A significant portion of this fuel was consumed immediately in the ensuing fireballs. The remaining fuel is believed either to have flowed down through the buildings or to have burned off within a few minutes of the aircraft impact. The heat produced by this burning jet fuel does not by itself appear to have been sufficient to initiate the structural collapses. However, as the burning jet fuel spread across several floors of the buildings, it ignited much of the buildings' contents, causing simultaneous fires across several floors of both buildings. The heat output from these fires is estimated to have been comparable to the power produced by a large commercial power generating station. Over a period of many minutes, this heat induced additional stresses into the damaged structural frames while simultaneously softening and weakening these frames. This additional loading and the resulting damage were sufficient to induce the collapse of both structures.

The ability of the two towers to withstand aircraft impacts without immediate collapse was a direct function of their design and construction characteristics, as was the vulnerability of the two towers to collapse a result of the combined effects of the impacts and ensuing fires. Many buildings with other design and construction characteristics would have been more vulnerable to collapse in these events than the two towers, and few may have been less vulnerable. It was not the purpose of this study to assess the code-conformance of the building design and construction, or to judge the adequacy of these features. However, during the course of this study, the structural and fire protection features of the buildings were examined. The study did not reveal any specific structural features that would be regarded as substandard, and, in fact, many structural and fire protection features of the design and construction were found to be superior to the minimum code requirements.

Several building design features have been identified as key to the buildings' ability to remain standing as long as they did and to allow the evacuation of most building occupants. These included the following:

- robustness and redundancy of the steel framing system

- adequate egress stairways that were well marked and lighted

- conscientious implementation of emergency exiting training programs for building tenants

Similarly, several design features have been identified that may have played a role in allowing the buildings to collapse in the manner that they did and in the inability of victims at and above the impact floors to safely exit. These features should not be regarded either as design deficiencies or as features that should be prohibited in future building codes. Rather, these are features that should be subjected to more detailed evaluation, in order to understand their contribution to the performance of these buildings and how they may perform in other buildings. These include the following:

- the type of steel floor truss system present in these buildings and their structural robustness and redundancy when compared to other structural systems

- use of impact-resistant enclosures around egress paths

- resistance of passive fire protection to blasts and impacts in buildings designed to provide resistance to such hazards

- grouping emergency egress stairways in the central building core, as opposed to dispersing them throughout the structure

During the course of this study, the question of whether building codes should be changed in some way to make future buildings more resistant to such attacks was frequently explored. Depending on the size of the aircraft, it may not be technically feasible to develop design provisions that would enable all structures to be designed and constructed to resist the effects of impacts by rapidly moving aircraft, and the ensuing fires, without collapse. In addition, the cost of constructing such structures might be so large as to make this type of design intent practically infeasible.

Although the attacks on the World Trade Center are a reason to question design philosophies, the BPS Team believes there are insufficient data to determine whether there is a reasonable threat of attacks on specific buildings to recommend inclusion of such requirements in building codes. Some believe the likelihood of such attacks on any specific building is deemed sufficiently low to not be considered at all. However, individual building developers may wish to consider design provisions for improving redundancy and robustness for such unforeseen events, particularly for structures that, by nature of their design or occupancy, may be especially susceptible to such incidents. Although some conceptual changes to the building codes that could make buildings more resistant to fire or impact damage or more conducive to occupant egress were identified in the course of this study, the BPS Team felt that extensive technical, policy, and economic study of these concepts should be performed before any specific code change recommendations are developed. This report specifically recommends such additional studies. Future building code revisions may be considered after the technical details of the collapses and other building responses to damage are better understood.

Several other buildings, including the Marriott Hotel (WTC 3), the South Plaza building (WTC 4), the U.S. Customs building (WTC 6), and the Winter Garden, experienced severe damage as a result of the massive quantities of debris that fell on them when the two towers collapsed. The St. Nicholas Greek Orthodox Church just south of WTC 2 was completely destroyed by the debris that fell on it.

Executive Summary

WTC 5, WTC 7, 90 West Street, the Bankers Trust building, the Verizon building, and World Financial Center 3 were impacted by large debris from the collapsing towers and suffered structural damage, but arrested collapse to localized areas. The performance of these buildings demonstrates the inherent ability of redundant steel-framed structures to withstand extensive damage from earthquakes, blasts, and other extreme events without progressive collapse.

The debris from the collapses of the WTC towers also initiated fires in surrounding buildings, including WTC 4, 5, 6, and 7; 90 West Street; and 130 Cedar Street. Many of the buildings suffered severe fire damage but remained standing. However, two steel-framed structures experienced fire-induced collapse. WTC 7 collapsed completely after burning unchecked for approximately 7 hours, and a partial collapse occurred in an interior section of WTC 5. Studies of WTC 7 indicate that the collapse began in the lower stories, either through failure of major load transfer members located above an electrical substation structure or in columns in the stories above the transfer structure. The collapse of WTC 7 caused damage to the Verizon building and 30 West Broadway. The partial collapse of WTC 5 was not initiated by debris and is possibly a result of fire-induced connection failures. The collapse of these structures is particularly significant in that, prior to these events, no protected steel-frame structure, the most common form of large commercial construction in the United States, had ever experienced a fire-induced collapse. Thus, these events may highlight new building vulnerabilities, not previously believed to exist.

In the study of the WTC towers and the surrounding buildings that were subsequently damaged by falling debris and fire, several issues were found to be critical to the observed building performance in one or more buildings.

These issues fall into several broad topics that should be considered for buildings that are being evaluated or designed for extreme events. It may be that some of these issues should be considered for all buildings; however, additional studies are required before general recommendations, if any, can be made for all buildings. The issues identified from this study of damaged buildings in or near the WTC site have been summarized into the following points:

a. Structural framing systems need redundancy and/or robustness, so that alternative paths or additional capacity are available for transmitting loads when building damage occurs.

b. Fireproofing needs to adhere under impact and fire conditions that deform steel members, so that the coatings remain on the steel and provide the intended protection.

c. Connection performance under impact loads and during fire loads needs to be analytically understood and quantified for improved design capabilities and performance as critical components in structural frames.

d. Fire protection ratings that include the use of sprinklers in buildings require a reliable and redundant water supply. If the water supply is interrupted, the assumed fire protection is greatly reduced.

e. Egress systems currently in use should be evaluated for redundancy and robustness in providing egress when building damage occurs, including the issues of transfer floors, stair spacing and locations, and stairwell enclosure impact resistance.

f. Fire protection ratings and safety factors for structural transfer systems should be evaluated for their adequacy relative to the role of transfer systems in building stability.

The BPS Team has developed recommendations for specific issues, based on the study of the performance of the WTC towers and surrounding buildings in response to the impact and fire damage that occurred. These recommendations have a broader scope than the important issue of building concepts and design for mitigating damage from terrorist attacks, and also address the level at which resources should be expended for aircraft security, how the fire protection and structural engineering communities should

Executive Summary

increase their interaction in building design and construction, possible considerations for improved egress in damaged structures, the public understanding of typical building design capacities, issues related to the study process and future activities, and issues for communities to consider when they are developing emergency response plans that include engineering response.

National Response. Resources should be directed primarily to aviation and other security measures rather than to hardening buildings against airplane impact. The relationship and cooperation between public and private organizations should be evaluated to determine the most effective mechanisms and approaches in the response of the nation to such disasters.

Interaction of Structural Elements and Fire. The existing prescriptive fire resistance rating method (ASTM E119) does not provide sufficient information to determine how long a building component in a structural system can be expected to perform in an actual fire. A method of assessing performance of structural members and connections as part of a structural system in building fires is needed for designers and emergency personnel.

The behavior of the structural system under fire conditions should be considered as an integral part of the structural design. Recommendations are to:

- Develop design tools, including an integrated model that predicts heating conditions produced by the fire, temperature rise of the structural component, and structural response.

- Provide interdisciplinary training in structures and fire protection for both structural engineers and fire protection engineers.

Performance criteria and test methods for fireproofing materials relative to their durability, adhesion, and cohesion when exposed to abrasion, shock, vibration, rapid temperature rise, and high-temperature exposures need further study.

Interaction of Professions in Design. The structural, fire protection, mechanical, architectural, blast, explosion, earthquake, and wind engineering communities need to work together to develop guidance for vulnerability assessment, retrofit, and the design of concrete and steel structures to mitigate or reduce the probability of progressive collapse under single- and multiple-hazard scenarios.

An improved level of interaction between structural and fire protection engineers is encouraged. Recommendations are to:

- Consider behavior of the structural system under fire as an integral part of the design process.

- Provide cross-training of fire protection and structural engineers in the performance of structures and building fires.

Fire Protection Engineering Discipline. The continued development of a system for performance-based design is encouraged. Recommendations are to:

- Improve the existing models that simulate fire and spread in structures, as well as the impact of fire and smoke on structures and people.

- Improve the database on material burning behavior.

Building Evacuation. The following topics were not explicitly examined during this study, but are recognized as important aspects of designing buildings for impact and fire events. Recommendations for further study are to:

- Perform an analysis of occupant behavior during evacuation of the buildings at WTC to improve the design of fire alarm and egress systems in high-rise buildings.

Executive Summary

- Perform an analysis of the design basis of evacuation systems in high-rise buildings to assess the adequacy of the current design practice, which relies on phased evacuation.

- Evaluate the use of elevators as part of the means of egress for mobility-impaired people as well as the general building population for the evacuation of high-rise buildings. In addition, the use of elevators for access by emergency personnel needs to be evaluated.

Emergency Personnel. One of the most serious dangers firefighters and other emergency responders face is partial or total collapse of buildings. Recommended steps to provide better protection to emergency personnel are to:

- Have fire protection and structural engineers assist emergency personnel in developing pre-plans for buildings and structures to include more detailed assessments of hazards and response of structural elements and performance of buildings during fires, including identification of critical structural elements.

- Develop training materials and courses for emergency personnel concerning the effects of fire on steel.

- Review collaboration efforts between the emergency personnel and engineering professions so that engineers may assist emergency personnel in assessments during an incident.

Education of Stakeholders. Stakeholders (e.g., owners, operators, tenants, authorities, designers) should be further educated about building codes, the minimum design loads typically addressed for building design, and the extreme events that are not addressed by building codes. Should stakeholders desire to address events not included in the building codes, they should understand the process of developing and implementing strategies to mitigate damage from extreme events.

Stakeholders should also be educated about the expected performance of their building when renovations, or changes in use or occupancy, occur and the building is subjected to different floor or fire loads. For instance, if the occupancy in a building changes to one with a higher fire hazard, stakeholders should have the fire protection systems reviewed to ensure there is adequate fire protection. Or, if the structural load is increased with a new occupancy, the structural support system should be reviewed to ensure it can carry the new load.

Study Process. This report benefited from a tremendous amount of professional volunteerism in response to this unprecedented national disaster. Improvements can be made that would aid the process for any future efforts. Recommendations are to:

- Provide resources that are proportional to the required level of effort.

- Provide better access to data, including building information, interviews, samples, site photos, and documentation.

Archival Information. Archival information has been collected and provides the groundwork for continued study. It is recommended that a coordinated effort for the preservation of this and other relevant information be undertaken by a responsible organization or agency, capable of maintaining and managing such information. This effort would include:

- cataloging all photographic data collected to date

- enhancing video data collected for both quality and timeline

- conducting interviews with building occupants, witnesses, rescue workers, and any others who may provide valuable information

- initiating public requests for information

Executive Summary

SEAoNY Structural Engineering Emergency Response Plan. As with any first-time event, difficulties were encountered at the beginning of the relationship between the volunteer engineering community and the local government agencies. Lessons learned in hindsight can be valuable to other engineering and professional organizations throughout the country. Appendix F presents recommendations that can be used as a basis for the development of other, similar plans.

Therese McAllister
Jonathan Barnett
John Gross
Ronald Hamburger
Jon Magnusson

1 *Introduction*

1.1 Purpose and Scope of Study

The events in New York City (NYC) on September 11, 2001, were among the worst building disasters and loss of life from any single building event in the United States. A total of 2,830 people lost their lives that day at the World Trade Center (WTC) site, including 403 emergency responders. The nation was shocked by the attacks and resulting collapse of office buildings that had been in use every day.

This report presents observations, findings, and recommendations regarding the performance of buildings affected by the September 11 attacks on the WTC towers in New York City. This report also describes the structural and fire protection features of the affected buildings and their performance in response to the terrorist attacks. Due to the unprecedented nature, magnitude, and visibility of the terrorist attacks, this event is among the most well-documented in the media, particularly in terms of photographic images, lives affected, and the immediate responses and ensuing sequence of events. An understanding of these events must include the performance of the buildings under extreme conditions beyond building code requirements. This includes determining the probable causes of collapse and identifying lessons to be learned. Recommendations are presented for more detailed engineering studies, to complete the assessments and to produce improved guidance for building design and performance evaluation tools.

During the September 11 attacks, a large number of buildings were extensively damaged by impact and fire events. To study the response of the affected buildings, a diverse group of experts in tall building design, steel structure behavior, fire protection engineering, blast effects, and structural investigations was empaneled into a Building Performance Study (BPS) Team. The study was sponsored by the Federal Emergency Management Agency (FEMA) and the Structural Engineering Institute of the American Society of Civil Engineers (SEI/ASCE). In conducting the study, the BPS Team received tremendous cooperation from the State of New York, the New York City Department of Design and Construction (DDC), the New York City Office of Emergency Management (OEM), the Port Authority of New York and New Jersey (hereafter referred to as the Port Authority), the National Institute of Standards and Technology (NIST), and the Structural Engineers Association of New York (SEAoNY). In addition, the BPS Team was supported by a coalition of organizations that included the American Concrete Institute (ACI), the American Institute of Steel Construction (AISC), the Council of American Structural Engineers (CASE), the International Code Council (ICC), the Council on Tall Buildings and Urban Habitat (CTBUH), the National Council of Structural Engineers Associations (NCSEA), the National Fire Protection Association (NFPA), the Society of Fire Protection Engineers (SFPE), and the Masonry Society (TMS).

FEMA and ASCE began discussing site studies and teams on September 12, as engineers and emergency management agencies all over the nation rallied to provide support. A number of the team members were at the site immediately after the attacks to assist as needed. As soon as the rescue operations were halted and the

CHAPTER 1: *Introduction*

FEMA Urban Search and Rescue teams left the site, the BPS Team mobilized to the WTC site and conducted field observations during the week of October 7, 2001. While in New York, the team inspected and photographed the site and individual building conditions, visited the salvage yards receiving steel from the collapsed buildings, attended presentations by design professionals associated with WTC buildings, and reviewed available building drawings. Upon completion of the site visit, the team members continued to collect extensive information and data, including photographs, video footage, and emergency response radio communications; continued surveillance of steel delivered to the recycling yards with the support of SEAoNY volunteers; and conducted additional interviews with direct witnesses of the events as well as participants in the original building design, construction, and maintenance. This information led to the development of a timeline of building loading events and allowed an initial engineering assessment of building performance. The study focus was to determine probable failure mechanisms and to identify areas of future investigation that could lead to practical measures for improving the damage resistance of buildings against such unforeseen events.

1.2 WTC Site

The World Trade Center and adjacent affected buildings were located on New York City's lower west side, adjacent to the Hudson River at the southern tip of Manhattan. As shown in Figure 1-1, the WTC site itself comprises 16 acres with buildings grouped around a 5-acre plaza. It is bounded by Vesey Street to the north, Church Street to the east, Liberty Street to the south, and West Street to the west. The WTC Complex consisted of seven buildings (referred to in this report as WTC 1 through WTC 7), the Port Authority Trans-Hudson (PATH) and Metropolitan Transit Authority (MTA) WTC stations, and associated Concourse areas. The WTC Plaza and its six buildings were originally developed by the Port Authority. Groundbreaking for construction was on August 5, 1966. Steel erection began in August 1968. First tenant occupancy of the 110-story north tower (WTC 1) was in December 1970, and occupancy of the 110-story south tower (WTC 2) began in January 1972. The other WTC buildings were constructed during the 1970s and into the 1980s, with WTC 7 constructed just north of the WTC site in 1985. WTC 3, located immediately west of the south tower, was a 22-story hotel operated by the Marriott Corporation. WTC 4 and 5 were nine-story office buildings, and WTC 6 was an eight-story office building. WTC 7 was a 47-story office building. The seven-building complex provided approximately 12 million square feet of rentable floor space occupied by a variety of government and commercial tenants. Many of the commercial tenants were in the insurance and financial industries. At the time of the September 11 attacks, the entire project had been transferred to a private party under a 99-year capital lease.

The New York Stock Exchange and the Wall Street financial district are located about three blocks southeast of the site. The World Financial Center (WFC) complex was constructed in the early 1980s and is located directly to the west, across West Street. Other prominent buildings immediately surrounding the WTC site include a historic Cass Gilbert designed building at 90 West Street and the Bankers Trust building at 130 Liberty Street, both located immediately to the south; the 1 Liberty Plaza building, located to the east; and the Verizon building, located directly to the north.

A six-story subterranean structure was underneath a large portion of the main WTC Plaza and WTC 1, 2, 3, and 6. Material excavated to construct this site was used to fill a portion of the Hudson River shoreline just across West Street and to create the adjacent World Financial Center (WFC) site. Construction of this deep substructure was a significant challenge, given the proximity of the Hudson River and the presence of a number of tall buildings along the south, east, and north sides of the site. In order to aid the excavation, slurry wall technology was utilized. In this technology, a trench is dug in the eventual location of the perimeter retaining walls. A bentonite slurry is pumped into the trench as it is excavated, and used to keep the trench open against the surrounding earth. Reinforcing steel is lowered into the trench, and

CHAPTER 1: *Introduction*

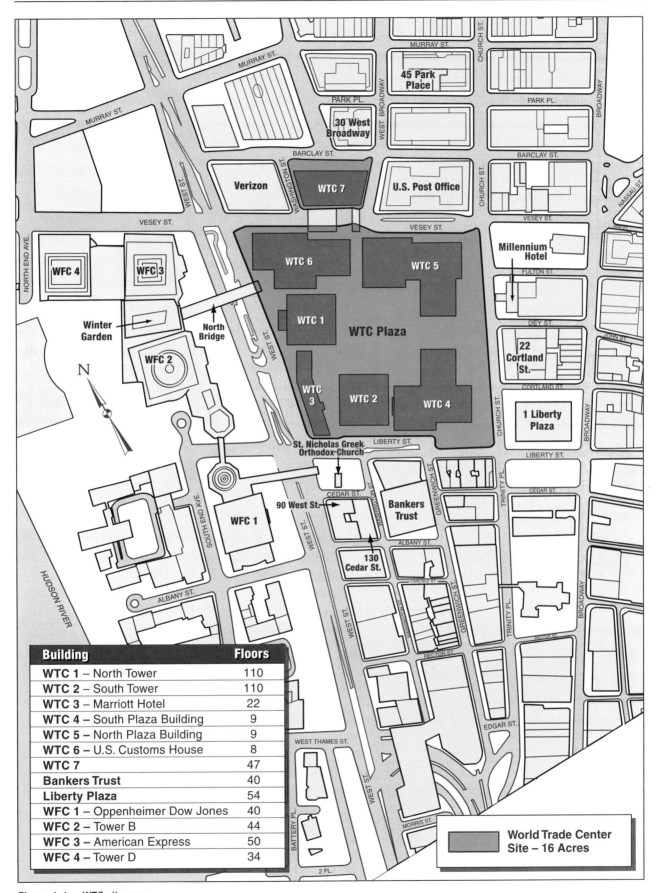

Figure 1-1   WTC site map.

concrete is placed through a tremie to create a reinforced concrete wall around the site perimeter. After the concrete wall is cured, excavation of the substructure begins. As the excavation progresses below surrounding grade, tiebacks are drilled through the exposed concrete wall and through the surrounding soil into the rock below to provide stability for the excavation. At the WTC site, these tiebacks were temporary and were replaced in the final construction by the subterranean floor slabs that provided lateral support to the walls.

A further challenge to the construction of the substructure was the presence of two existing subway lines across the site. The Interboro Rapid Transit System 1 and 9 subway lines, operated by the MTA, ran north to south across the middle of the site adjacent to the east wall of the substructure. A second subway system, PATH, operated by the Port Authority, made a 180-degree terminal bend beneath the western half of the site. This subway tunnel was temporarily supported across the excavation and incorporated into the final construction with a station provided for this line inside the slurry wall, just west of the 1 and 9 subway lines, and below the Plaza area just east of WTC 1 and partially below WTC 6. Although significant damage was sustained by the buildings, subterranean structure, and subway system, only the performances of the above-grade buildings were assessed in this study.

1.3 Timeline and Event Summary

On the morning of September 11, 2001, two hijacked commercial jetliners were deliberately flown into the WTC towers. The first plane, American Airlines Flight 11, originated at Boston's Logan International Airport at 7:59 a.m., Eastern Daylight Time. The plane was flown south, over midtown Manhattan, and crashed into the north face of the north tower (WTC 1) at 8:46 a.m. The second plane, United Airlines Flight 175, departed Boston at 8:14 a.m., and was flown over Staten Island and crashed into the south face of the south tower (WTC 2) at 9:03 a.m.

Both flights, scheduled to arrive in Los Angeles, were Boeing 767-200ER series aircraft loaded with sufficient fuel for the transcontinental flights. These aircraft are described in Appendix E. There were 92 people on board Flight 11 and 65 people on board Flight 175. Figure 1-2 shows the approximate flight paths for the two aircraft.

The north tower was struck between floors 94 and 98, with the impact roughly centered on the north face. The south tower was hit between floors 78 and 84 toward the east side of the south face (Figures 1-3 and 1-4). Each plane banked steeply as it was flown into the building, causing damage across multiple floors. According to Government sources, the speed of impact into the north tower was estimated to be 410 knots, or 470 miles per hour (mph), and the speed of impact into the south tower was estimated to be 510 knots, or 590 mph. As the two aircraft impacted the buildings, fireballs erupted (Figure 1-5) and jet fuel spread across the impact floors and down interior shaftways, igniting fires (Figure 1-6). The term fireball is used to describe deflagration, or ignition, of a fuel vapor cloud. As the resulting fires raged throughout the upper floors of the two WTC towers, thousands attempted to evacuate the buildings. It was estimated by the Port Authority that the population of the WTC complex on September 11, 2001, was 58,000 people. This estimate includes the PATH and MTA stations and the Concourse areas. Almost everyone in WTC 1 and WTC 2 who was below the impact areas was able to safely evacuate the buildings, due to the length of time between the impact and collapse of the individual towers.

At 9:59 a.m., 56 minutes after it was struck, the south tower collapsed. The north tower continued to stand until 10:29 a.m., when it, too, collapsed. The north tower had survived 1 hour and 43 minutes from the time the jetliner crashed into it. A total of 2,830 people lost their lives in the collapse of the WTC towers, including 2,270 building occupants, 157 airplane crew and passengers, and 403 firefighters, police personnel, and other emergency responders.

CHAPTER 1: *Introduction*

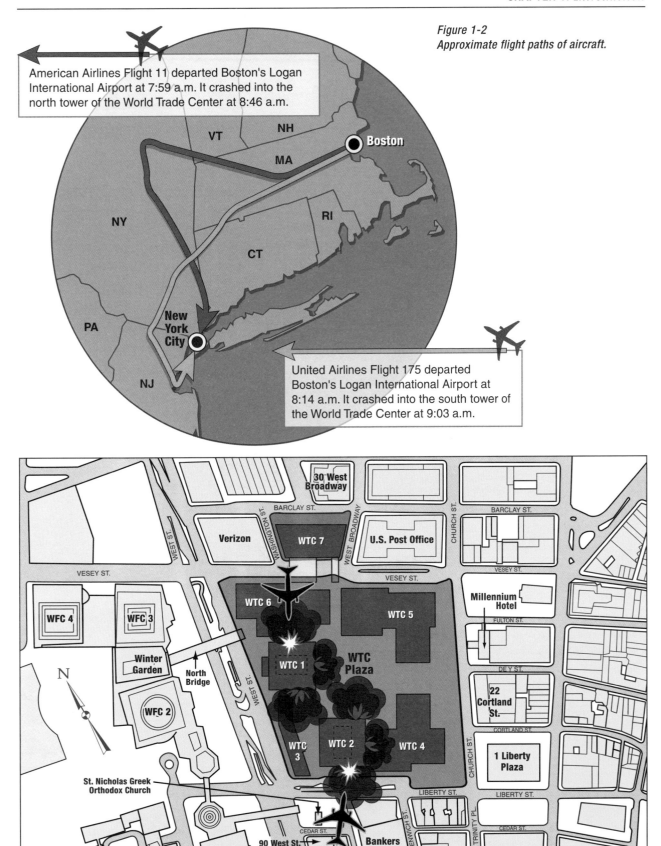

Figure 1-2
Approximate flight paths of aircraft.

American Airlines Flight 11 departed Boston's Logan International Airport at 7:59 a.m. It crashed into the north tower of the World Trade Center at 8:46 a.m.

United Airlines Flight 175 departed Boston's Logan International Airport at 8:14 a.m. It crashed into the south tower of the World Trade Center at 9:03 a.m.

Figure 1-3 *WTC impact locations and resulting fireballs.*

FEDERAL EMERGENCY MANAGEMENT AGENCY

CHAPTER 1: *Introduction*

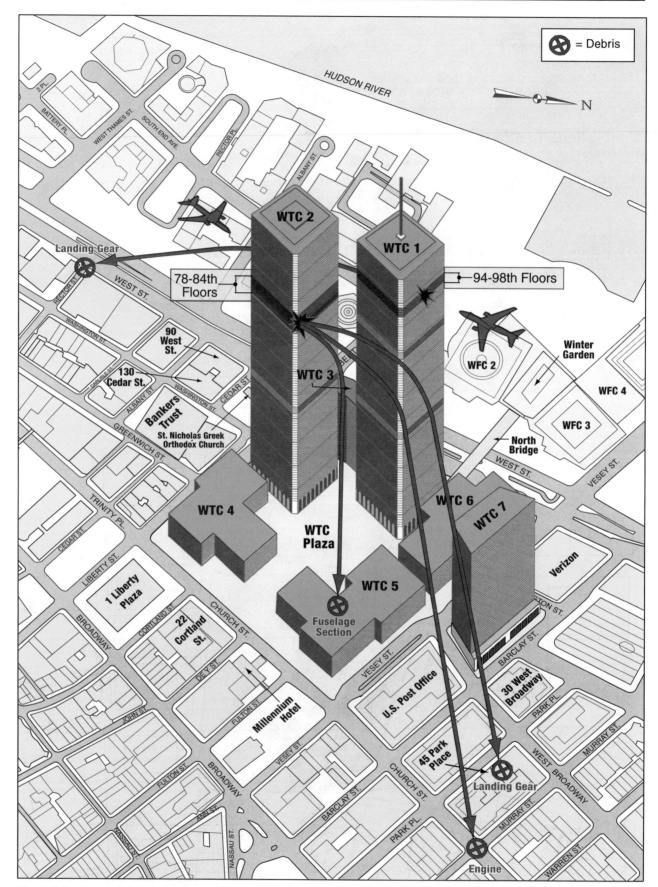

Figure 1-4   Areas of aircraft debris impact.

CHAPTER 1: *Introduction*

Figure 1-5   Fireball erupts on the north face of WTC 2 as United Airlines Flight 175 strikes the building.

Figure 1-6   View of the north and east faces showing fire and impact damage to both towers.

FEDERAL EMERGENCY MANAGEMENT AGENCY

CHAPTER 1: *Introduction*

Debris from the collapsing towers, some of it still on fire, rained down on surrounding buildings, causing structural damage and starting new fires (Figure 1-7). The sudden collapse of each tower sent out air pressure waves that spread dust clouds of building materials in all directions for many blocks. The density and pressure of the dust clouds were strong enough to carry light debris and lift or move small vehicles and break windows in adjacent buildings for several blocks around the WTC site. Most of the fires went unattended as efforts were devoted to rescuing those trapped in the collapsed towers. The 22-story Marriott World Trade Center Hotel (WTC 3) was hit by a substantial amount of debris during both tower collapses. Portions of WTC 3 were severely damaged by debris from each tower collapse, but progressive collapse of the building did not occur. However, little of WTC 3 remained standing after the collapse of WTC 1. WTC 4, 5, and 6 had floor contents and furnishings burn completely and suffered significant partial collapses from debris impacts and from fire damage to their structural frames. WTC 7, a 47-story building that was part of the WTC complex, burned unattended for 7 hours before collapsing at 5:20 p.m. The falling debris also damaged water mains around the WTC site at the following locations:

- 20-inch main on West Street, closed to the slurry wall, about midway between Vesey Street and Liberty Street
- 20-inch main along the Financial Center north of the South Link Bridge
- 20-inch main at the corner of Liberty Street and West Street
- main in front of the West Street entrance to 90 West
- 24-inch main on Vesey Street, near West Street
- main at the corner of Vesey Street and West Broadway, near the subway station
- main at the southwest edge of 30 West Broadway
- 16-inch main inside the slurry wall

Damaged mains were located after the collapses, but access was impeded by the collapse debris.

The timeline of the major events is summarized in Table 1.1. The times and seismic data were recorded at the Lamont-Doherty Earth Observatory (LDEO) of Columbia University. The signal duration and Richter Scale magnitudes were included to indicate the relative magnitudes of energy transmitted through the ground between the events. Figure 1-8 shows the accelograms recorded by the observatory during the events.

Other buildings surrounding the WTC plaza were also damaged by falling debris. A few buildings, such as the Bankers Trust building, suffered significant damage but remained standing. Many buildings had their façades and glazing damaged and their interiors blanketed with debris from the collapse of the WTC towers and WTC 7. Figures 1-9A and 1-9B are satellite images of the WTC site taken before and after the September 11 attacks, respectively.

1.4 Response of the Engineering Community

1.4.1 Local Authorities

Immediately after the attacks, it became apparent to the City of New York that there was an enormous need for structural engineering and construction expertise and support. Within hours, the DDC appealed to several construction companies (Bovis/Lend-Lease, AMEC, Turner-Plaza and Tully) and the engineering firm, LZA Technology/Thornton-Tomasetti (LZA) to assist in the search and rescue effort. Mobilization began immediately. A reconnaissance inspection by DDC and LZA took place in the afternoon of September 11. A first round of building inspections was performed on September 12 by engineers from DDC, the NYC Department of Buildings (DoB), and LZA.

CHAPTER 1: *Introduction*

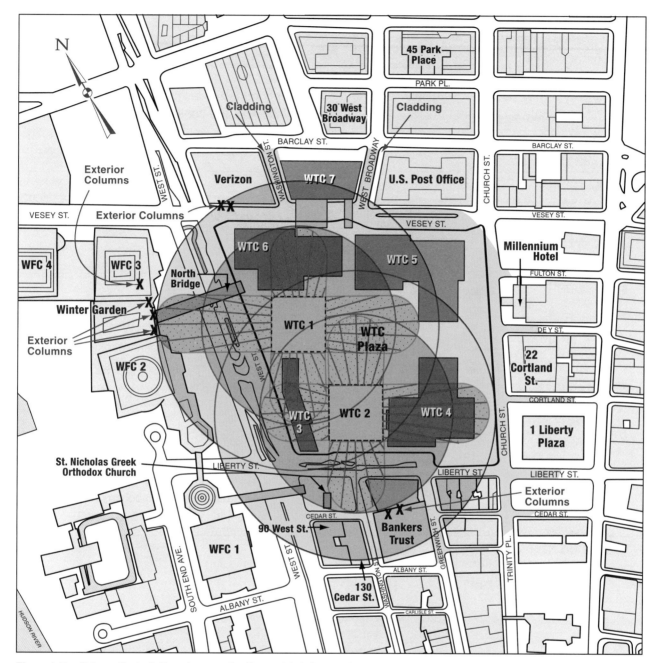

Figure 1-7 Schematic depiction of areas of collapse debris impact, based on aerial photographs and documented damage. Striped areas indicate predominant locations of exterior steel columns. Inner circles indicate approximate radius of exterior steel columns and other heavy debris. Outer circles indicate approximate radius of aluminum cladding and other lighter debris. Heavy Xs show where exterior steel columns were found outside the predominate debris areas.

DDC, the agency that had responsibility to manage all construction and engineering at the site, was joined by engineers and construction managers from the Port Authority on the following days. Beginning on September 13, consulting support was provided by SEAoNY, Mueser Rutledge Consulting Engineers, Leslie E. Robertson Associates, the U.S. Army Corps of Engineers, FEMA Urban Search and Rescue, and various other New York City departments.

The engineering efforts had two objectives – safety of personnel involved in the recovery process and evaluation of the conditions and processes that would allow a return to safe occupancy of the buildings in the area.

CHAPTER 1: *Introduction*

Table 1.1 Timeline of Major Events[1]

Start Time[2]	Signal Duration	Magnitude (Richter Scale)	Event
8:46:26 EDT (12:46 UTC)	12 seconds	0.9	WTC 1 (the north tower) was hit by American Airlines Flight 11, a hijacked 767-200ER commercial jet airliner.
9:02:54 EDT (13:02 UTC)	6 seconds	0.7	WTC 2 (the south tower) was hit by United Airlines Flight 175, also a hijacked 767-200ER jet.
9:59:04 EDT (13:59 UTC)	10 seconds	2.1	WTC 2 began collapsing after 56 minutes, 10 seconds. Large debris from the collapse fell on WTC 3 and WTC 4, 130 Cedar Street, 90 West Street, and Bankers Trust. WTC 3 suffered a partial collapse. Fire was initiated in WTC 4 and 90 West Street.
10:28:31 EDT (14:28 UTC)	8 seconds	2.3	WTC 1 began collapsing after 102 minutes, 5 seconds. Large debris from the collapse fell on WTC 3, 5, 6, and 7; the Winter Garden; and the American Express (World Financial Center 2) building. WTC 3 collapsed to the 3rd floor, and fires were initiated in WTC 5, 6, and 7.
17:20:33 EDT (21:20 UTC)	18 seconds	0.6	WTC 7 began collapsing.

[1] Based on seismic recordings made by the Lamont-Doherty Earth Observatory of Columbia University, 34 kilometers north of the WTC site.

[2] EDT = Eastern Daylight Time; UTC = Coordinated Universal Time. Times cited in this report are based on these times, rounded to the nearest minute.

1.4.2 SEAoNY Participation

Immediately after the attacks, members of the Board of Directors of SEAoNY initiated contact with DDC, DoB, and OEM. By Wednesday morning, September 12, the Board had established communications with the New York Police Department (NYPD), OEM, and DDC. SEAoNY teams of structural engineers were retained through their firms by LZA and began assisting with the rescue and recovery efforts on Thursday, September 13. They served continuously (24 hours a day, 7 days a week) through January 9, 2002. The SEAoNY teams provided engineering guidance with search and rescue, demolition, and temporary construction, as well as assistance to contractors working to stabilize or remove debris.

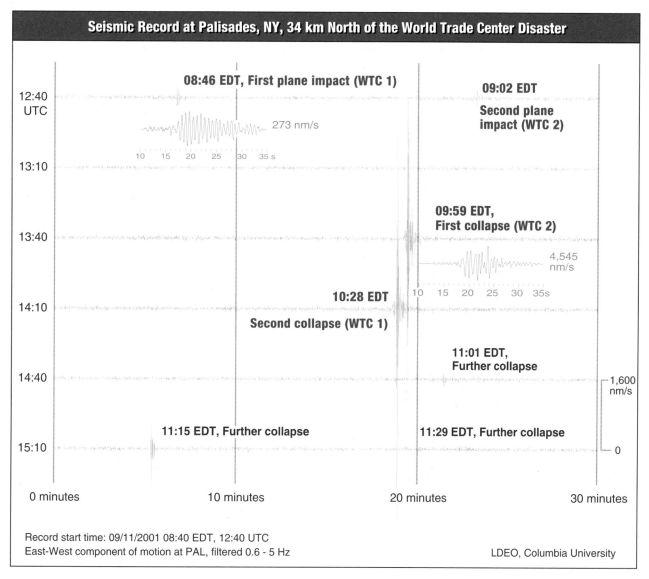

Figure 1-8 Seismic recordings on east-west component at Palisades, NY, for events at WTC on September 11, 2001, distance 34 km. Three hours of continuous data are shown starting at 08:40 EDT (12:40 UTC). The two largest signals were generated by the collapses of towers 1 and 2. Expanded views of first impact and first collapse are shown in red. The amplitude of the seismic signal is in nanometers per second (nm/s), and the peak amplitude of the ground motion at this station reached to 4,545 nm/s for the first collapse. Note the relatively periodic motions for impacts 1 and 2.

Structural engineers acted as guides through the site, providing descriptions of structures and offering judgment on the stability of structures and debris. They provided warnings of potential hazards and assisted with choosing crane and other equipment locations and installations. Assisting the LZA team, structural engineers worked in four teams staffed by SEAoNY members. In the first 30 days after the attack, more than 10,000 engineer hours, or 1-1/2 engineer-years per week, were expended by these teams.

DDC was also asked on September 14 to rapidly assess the condition of the more than 400 buildings in lower Manhattan suspected of being damaged by the collapse of the WTC towers. It appeared that the zone of damage was significant and that many buildings may have received debris or vibration damage. The potentially hazardous conditions would make the area unsafe for rescue and removal personnel. DDC and LZA assigned this task to SEAoNY. These systematic building inspections were organized, coordinated, and

CHAPTER 1: *Introduction*

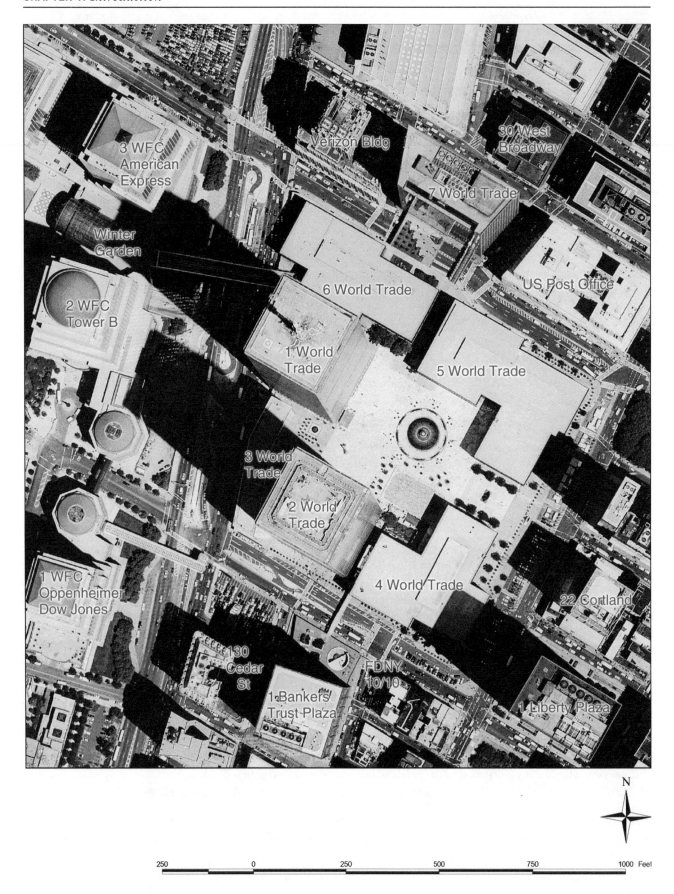

Figure 1-9A   Satellite photograph of the WTC site taken before the attacks.

Figure 1-9B Satellite photograph of the WTC site taken after the attacks.

CHAPTER 1: *Introduction*

performed by members of SEAoNY. Engineering firms representing those members worked as consultants to LZA. The purpose of the building assessment was to assist in determining which buildings could be safely reoccupied and to identify structural or falling hazards that might injure site personnel or the public.

Similar efforts had been conducted by Structural Engineers Associations in the western United States following earthquakes; however, no formal mechanisms were in place in New York for this type of effort. The field manual *ATC 20, Procedures for Post-earthquake Safety Evaluation of Buildings,* published by the Applied Technology Council (ATC) was used for procedural guidance on performing the inspections. Custom forms for rapid visual assessments were created. Additional teams of structural engineers organized by SEAoNY completed the first round of assessments of approximately 400 buildings on September 17 and 18.

After the initial assessments had been completed, additional inspections were recommended for the buildings that appeared to be the most distressed. These additional inspections were completed within a matter of days after the recommendations were discussed with DDC. As a follow-up, a second round of evaluations for all of the buildings was performed between October 4 and 10, and detailed engineering reports for the most severely damaged buildings (outside of the WTC site) were prepared in conjunction with DDC and LZA. Building evaluation summaries are presented in Chapter 7 (Peripheral Buildings).

In the initial response, most of SEAoNY's members volunteered, as did structural engineers from across the country who coordinated with SEAoNY through NCSEA. SEAoNY members located in New York City also volunteered office space, equipment, and support to other engineers working at the WTC site. To eliminate any potential liability issues and to meet the long-term commitment required by the engineers due to the magnitude of the event, all of these engineers were retained by DDC through LZA.

SEAoNY volunteers provided assistance to the BPS Team by maintaining five teams of structural engineers to monitor steel debris from the WTC site as it arrived at the salvage yards. Their goal was to locate material from the impact zones. SEAoNY engineers also collected hundreds of hours of video and thousands of still photographs from other engineers and the general public in an effort to fully document the collapses and the recovery operations.

As with any first-time event, difficulties were encountered at the beginning of the relationship between the volunteer engineering community and the local government agencies. There were no procedures for either the engineers or the agencies to follow for such an event, leading to a situation in which the organization of the work and the procedures to be followed were developed and revised almost daily in response to the circumstances.

Issues of identification, credentials, responsibility, and liability required considerable attention. Because the site was treated as a crime scene, access at various locations had to be obtained through checkpoints manned by the National Guard, the Fire Department of New York (FDNY), and the NYPD. Also, because there was no identification system in place for the first few days, it took up to 3 hours for SEAoNY volunteers to get to the command center from the outer perimeter of the site, a distance of less than six blocks. In addition, there were issues related to the responsibilities and liabilities of individual volunteers, their firms, and SEAoNY. Currently in New York City, there is no process for deputizing volunteers, nor are there any "Good Samaritan" laws in effect.

Lessons learned in hindsight can be valuable to other engineering and professional organizations throughout the country. SEAoNY has drafted a "Structural Engineering Emergency Response Plan (SEERP)," which will be used in discussions with New York City to develop and formalize relationships and procedures that will improve responses to any future emergency events or disasters. Appendix F presents the draft SEERP, which may be useful to other cities considering such activities.

CHAPTER 1: *Introduction*

1.5 Overview of Building Codes and Fire Standards

1.5.1 Building Codes

Building design and occupancy in the United States is governed by building codes that specify the minimum environmental, or external, loads that a building must have the strength to resist. They also prescribe requirements for internal challenges such as fire protection and timely egress for life safety. However, designers must consider project circumstances and owner requirements when determining building design loads. The primary external loads specified are:

- gravity
- wind
- earthquake

Other risks considered include the potential for fire, hazardous material leaks or explosion, and the need to promptly evacuate occupants to safety. These demands establish building code requirements for fire resistive construction, emergency egress, and fire protection systems.

National model building codes do not include requirements to design for loads that might be imposed due to acts of war or terrorism. It is usually considered unnecessary to provide the capacity to resist such loads in most buildings; however, these loads may be included at the discretion of building owners if they desire a higher level of protection (e.g., an embassy, bank, or military facility).

Gravity loads include both the weight of the building and its contents. The weight of the building is calculated based on building construction plans and material densities. The weight of the contents is not specifically known at the time of design because it will depend on the building user and will vary with time. Therefore, the codes specify minimum floor loads on a pounds per square foot basis. For instance, for a standard office occupancy, the codes typically specify a minimum live load of 50 pounds per square foot (psf) of floor area. It is the responsibility of the building owners to see that floors are not overloaded.

Wind loads specified by codes are based on maps of design wind speed for different regions of the country. As wind speed increases, the wind pressure on the building increases proportionally to the square of the wind velocity. The pressure on the building also varies with the height and degree of shielding provided by other buildings and geographic features. Although not usually required by building codes, engineers frequently use wind tunnel studies to more accurately determine wind loads on tall buildings, where standard calculations may not be adequate. WTC 1, 2, 4, 5, and 6 all had extensive wind tunnel studies performed as part of the design process. WTC 1 and WTC 2 were among the first structures that were designed using wind tunnel studies.

The hazard presented by earthquakes is also highly dependent on the geographic region. In all regions of the country, including the most severe seismic areas of California, the effects of earthquakes are relatively small for very tall buildings. The flexibility of a very tall building, say 100 stories, generally allows the building to respond as the ground moves back and forth without developing forces nearly as large as those produced by design wind loads. Therefore, even in a severe seismic area, tall building design is generally controlled by wind loads.

Engineers design buildings for gravity, wind, and earthquake loads. Architects, in concert with fire protection professionals, including fire protection engineers, oversee the selection and application of fire resistive construction elements and fire protection systems such as sprinklers, fire alarms, and special hazard protection.

The building codes also prescribe the minimum requirements for life safety if a fire occurs. The prescribed fire safety features are intended to limit the fire threat and safeguard those in the building. Prime among these features are the egress and emergency notification (alarm) requirements.

Particularly in high-rise buildings, fire safety requirements are prescribed to maintain the integrity of the structure by controlling the intensity of fire and providing adequate structural strength during fires. In modern building codes, this is accomplished through a three-step approach:

- The first line of defense is the automatic sprinkler protection designed to control fires in their early stage of development and either extinguish them or hold the fire in check for the arrival of the fire department. Sprinklers are not normally capable of controlling fires that are of large size before the sprinklers can operate. Such was the case in the WTC towers.

- The second line of defense is manual firefighting by the fire department or fire brigades. The building is provided with standpipes, emergency control of elevators, special emergency communication systems, control centers, and other features to enable effective firefighting above the levels that can be attacked from the exterior. It was this line of defense that successfully controlled the 1975 fire. In the September 11 incident, the damage done to the elevators and the height of the fires precluded the fire department from being able to directly attack the fire. Even if they had reached the fire floors, they would have been faced with a fire situation possibly beyond even the excellent capabilities of the FDNY.

- The final line is the fire resistance of the building and its elements, including the building frame, the floors, partitions, shaft enclosures, and other elements that compartmentalize the building and structurally support it. Most important of these are the requirements for the structural frame (including columns, girders, and trusses). See Section 1.5.3 for a discussion of fire resistance ratings.

All three of these lines of defense were present, but were overwhelmed by the magnitude of the events of September 11, 2001.

For life safety and egress, stairways must have minimum widths based on the maximum number of occupants who may be in the building. Stairs must be separated from the remainder of the building by a minimum 1- or 2-hour fire resistant barrier to provide a level of safety while occupants traverse the stairs. At least two stairways must be provided with widely separated entry points. In most jurisdictions, elevators are designed to automatically return to the lobby level during a fire alarm to be controlled by firefighters. In many high-rise buildings, the elevator shafts and exit stairs are pressurized to keep out smoke and heat. The use of elevators is discouraged for emergency egress because of the potential for elevator failure and the likelihood of the elevator shaft acting like a chimney, carrying heat, smoke, and toxic gases throughout the building.

Fire protection systems (sprinklers, fire alarms, and special-hazard protection) are required to provide early notification and fire control until the fire department can arrive and begin manual suppression efforts. Smoke management systems are intended to aid emergency evacuation of building occupants and operations of emergency personnel. Manual suppression efforts by emergency personnel are aided by the presence of standpipe systems.

1.5.2 Unusual Building Loads

In planning a new building, an owner may request enhanced requirements in its design for events that are not anticipated by the building codes. In some cases, where unusual hazards such as explosive or toxic materials exist, the building codes prescribe special life safety and fire protection features. In most non-hazardous occupancies, these are not required. Only a very small percentage of buildings have extraordinary provisions for unusual circumstances and there is a limit to the events that can be handled and the strength capacities that can be provided. Defense facilities, nuclear power plants, and overseas embassies are just a few examples where special strengthening features are requested by building owners in the design and engineering of their facilities.

The WTC towers were the first structures outside of the military and the nuclear industries whose design considered the impact of a jet airliner, the Boeing 707. It was assumed in the 1960s design analysis for the WTC towers that an aircraft, lost in fog and seeking to land at a nearby airport, like the B-25 Mitchell bomber that struck the Empire State Building on July 28, 1945, might strike a WTC tower while low on fuel and at landing speeds. However, in the September 11 events, the Boeing 767-200ER aircraft that hit both towers were considerably larger with significantly higher weight, or mass, and traveling at substantially higher speeds. The Boeing 707 that was considered in the design of the towers was estimated to have a gross weight of 263,000 pounds and a flight speed of 180 mph as it approached an airport; the Boeing 767-200ER aircraft that were used to attack the towers had an estimated gross weight of 274,000 pounds and flight speeds of 470 to 590 mph upon impact.

Including aircraft impact as a design load requires selecting a design aircraft, as well as its speed, weight, fuel, and angle and elevation of impact. Figure 1-10 compares the design characteristics of several large aircraft that were in use or being planned for use during the life of the WTC towers. The maximum takeoff weight, fuel capacity, and cruise speed shown for each class of aircraft are presented for comparison of relative sizes and speeds. The larger square represents the floor plan area of the WTC towers (approximately 207 feet by 207 feet), and the smaller square represents a more typical size for a high-rise building. The likelihood of a building surviving an aircraft impact decreases as aircraft size and speed increase. The Airbus A380 is expected to be flying in 2006. Its weight and fuel capacity are approximately three times those of a 767-200ER. The security of aircraft is critical to the safety of high-rise and all other buildings; aircraft security measures should be commensurate with the size and potential risk posed by the aircraft.

The decision to include aircraft impact as a design parameter for a building would clearly result in a major change in the design, livability, usability, and cost of buildings. In addition, reliably designing a building to survive the impact of the largest aircraft available now or in the future may not be possible. These types of loads and analyses are not suitable for inclusion in minimum loads required for design of all buildings. Just as the possibility of a Boeing 707 impact was a consideration in the original design of WTC 1 and WTC 2, there may be situations where it is desirable to evaluate building survival for impact of an airplane of a specific size traveling at a specific speed. Although there is limited public information available on this topic (Bangash 1993, DOE 1996), interested building owners and design professionals would require further guidance for application to buildings.

1.5.3 Overview of Fire-Structure Interaction

Control of structural behavior under fire conditions has historically been based on highly prescriptive building code requirements. These requirements specify hourly fire resistance ratings. A popular misconception concerning fire resistance ratings for walls, columns, floors, and other building components is that the ratings imply the length of time that a building component will remain in place when exposed to an actual fire. For example, a 2-hour fire-resistant wall is often expected to remain standing for 2 hours if exposed to an actual fire. However, the time to collapse of such a wall in an actual fire may be greater or less than 2 hours. The standard method of test to evaluate fire resistance (ASTM E119) is a comparative test of relative specimen behavior under controlled conditions and is not intended to be predictive of actual behavior. Further, the results of the ASTM E119 test do not consider actual conditions such as member interactions, restraint, connections, or situations where damage to the structural assembly is present prior to initiation of the fire.

1.5.3.1 ASTM E119 Standard Fire Test

Building code requirements for structural fire protection are based on laboratory tests conducted in accordance with the Standard Test Methods for Fire Tests of Building Construction and Materials, ASTM E119 (also designated NFPA 251 and UL 263). Since its inception in 1918, the ASTM E119 Standard Fire Test has required that test specimens be representative of actual building construction. Achieving this requirement in actual practice has been difficult because available laboratory facilities can only accommodate floor specimens on the order of a 14-foot x 17-foot (4.3-meter x 5.2-meter) plan area in a fire test furnace. The specimens do not account for impact damage to fire protection coatings. For typical steel and concrete structural systems, the behavior of specimens in an ASTM E119 fire test does not reflect the behavior of floor and roof constructions that are exposed to uncontrolled fire in real buildings. The ASTM E119 fire endurance test exposes the test specimen to the time-temperature relationship shown in Appendix A, Figure A-9.

In contrast with the structural characteristics of ASTM E119 test specimens, floor slabs in real buildings are continuous over interior beams and girders, connections range from simple shear to full moment connections, and framing member size and geometry vary significantly, depending on structural system and building size and layout. Even for relatively simple structural systems, realistically simulating the restraint, continuity, and redundancy present in actual buildings is extremely difficult to achieve in a laboratory fire test assembly. In addition, the size and intensity of a real uncontrolled fire and the loads superimposed on a floor system during that exposure are variables not investigated during an ASTM E119 fire test. Many factors influence the intensity and duration of an uncontrolled fire and the likelihood of full design loads occurring simultaneously with peak fire temperatures is minimal.

The ASTM E119 Standard Fire Test was developed as a comparative test, not a predictive one. In effect, the Standard Fire Test is used to evaluate the relative performance (fire endurance) of different construction assemblies under controlled laboratory conditions, not to predict performance in real, uncontrolled fires.

1.5.3.2 Performance in Actual Building Fires

Extensive fire research in the United States and the international community established that the temperatures generated during an actual fire, represented by a time-temperature curve, is not only a function of the fire load, but also the following:

- ventilation (air access through the windows, doors, and heating, ventilation, and air conditioning [HVAC] system)
- compartment geometry (floor area, ceiling height, length to width to height ratios)
- thermal properties of the walls, floor, and ceiling construction
- combustion characteristics of the fuel (rate and duration of heat release)

International research in the past 30 years has substantiated the importance of ventilation rates. It is now recognized that two entirely different types of fires can occur within buildings or compartments. The first is a "fuel surface controlled fire" that will develop when compartment openings are sufficiently large to provide adequate combustion air for unrestricted burning. Such fires will generally be of short duration and the intensity will be controlled by the fire load and its arrangement.

The second type of fire is "ventilation controlled" and will develop when the compartment openings are not large enough to allow unrestricted burning. Such fires will burn longer than fires controlled by the amount of surface fuel. Fires in large spaces often burn in ventilation controlled and fuel surface controlled regimes, at different times during the fire and at different locations within the enclosure.

CHAPTER 1: *Introduction*

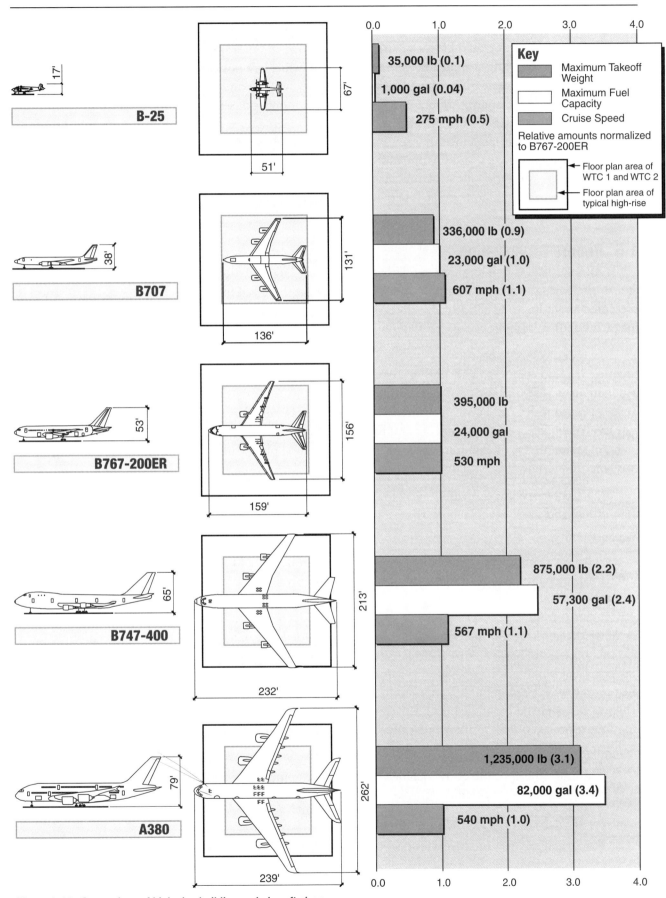

Figure 1-10 Comparison of high-rise building and aircraft sizes.

FEDERAL EMERGENCY MANAGEMENT AGENCY

1-19

These real fires contrast with building code requirements for fire resistant design, which are based on a presumed duration of a standard fire as a function of fire load and building occupancy, height, and area. The severity of actual fires is determined by additional factors, which are not typically considered in building codes except as an alternate material method or equivalency when accepted by the enforcing official (the authority having jurisdiction). Although there have been a number of severe fires in protected steel buildings, including the three described in Appendix A, Section A.3.1.3, the team is unaware of any protected steel structures that have collapsed in a fire prior to September 11. However, none of the other fire events had impact damage to structural and fire protection systems. Recent fire research provides a basis for designing more reliable fire protection for structural members by analytical methods that are becoming more acceptable to the building code community. Such methods were not available when the WTC buildings were designed in the 1960s.

1.6 Report Organization

All seven WTC buildings, the Bankers Trust building, and other buildings that sustained major impact and/or fire damage from the attacks on the WTC towers are discussed in detail in this report. Information is presented about building performance documented during this study, as well as findings and recommendations for each building, as appropriate.

In order to simultaneously conduct multiple investigations into each building, a Chapter Leader was assigned as a lead coordinator and author for each chapter or appendix. This approach allowed a high level of productivity and resulted in a different writing style for each chapter. Additionally, the scope and level of detail varies considerably between chapters. The two major factors that define chapter content are the type and level of damage a building suffered and the availability of building information during this study, including damage documentation, structural and architectural plans, fire protection systems, building contents, and modifications made during occupancy.

This report opens with an executive summary, followed by the Introduction (Chapter 1), which documents the purpose and scope of this report; the events and actions that occurred on September 11, 2001, at the WTC site; and background information on the building design and codes.

The WTC buildings are presented in Chapters 2 through 5, and are grouped by types of construction and damage. Chapter 2 presents observations, data, and the results of preliminary analyses conducted on each tower (WTC 1 and WTC 2). Chapter 3 briefly discusses the hotel (WTC 3) performance for two severe debris impact events from the collapsing towers. Chapter 4 includes WTC 4, 5, and 6, because all three buildings are of similar construction and experienced fire damage from debris. Chapter 5 presents observations, data, and preliminary analyses of WTC 7, which also suffered a complete collapse. Chapter 6 describes how the Bankers Trust building arrested a local collapse on the north side that was initiated by debris impact from the collapse of WTC 2. Buildings adjacent to the WTC site that sustained major damage are presented in Chapter 7 (Peripheral Buildings). Chapter 8 presents observations, findings, and recommendations for each building, as well as overall recommendations based on the collective assessment of individual building performance and related issues.

The following appendixes are included to allow development of pertinent issues and topics without interrupting the flow of the report:

A – Overview of Fire Protection in Buildings

B – Structural Steel and Steel Connections

C – Limited Metallurgical Examination

D – WTC Steel Data Collection

E – Aircraft Information

F – Structural Engineers Emergency Response Plan

G – Acknowledgments

H – Acronyms and Abbreviations

I – Metric Conversions

The reader should be aware that English units are the primary system of measurement in this report, except where temperature information is presented, such as in the discussion of fire protection systems. Temperatures are presented in degrees Celsius, followed by degrees Fahrenheit. This approach allows for ease of reading by general audiences while retaining the measurement system preferred by fire protection engineers.

1.7 References

American Society for Testing and Materials. 2000. *Standard Test Methods for Fire Tests of Building Construction and Materials*, ASTM E119. West Conshohocken, PA.

Baker, W. E.; Cox, P. A.; Westine, P. S.; Kulesz, J. J., Editors. 1982. *Explosion Hazards and Evaluation. Fundamental Studies in Engineering.* Elsevier Scientific Publishing Company, NY.

Bangash, M. 1993. *Impact and Explosion, Analysis and Design.* Section 4.3. CRC Press.

International Organization for Standardization. 1999. *Fire Resistance Tests – Elements of Building Construction*, ISO 834.

National Fire Protection Association. 1999. *Standard Methods of Fire Tests of Building Construction and Materials*, NFPA 251. Quincy, MA.

U.S. Department of Energy. 1996. *Accident Analysis for Aircraft Crash into Hazardous Facilities.* DOE Standard 3014-96. October.

Zalosh, R. G. 2002. "Explosion Protection," *SFPE Handbook of Fire Protection Engineering.* 3rd edition. Quincy, MA.

Ronald Hamburger
William Baker
Jonathan Barnett
Christopher Marrion
James Milke
Harold "Bud" Nelson

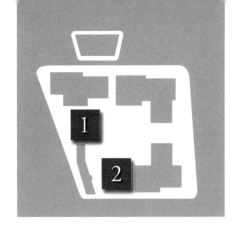

2 WTC 1 and WTC 2

2.1 Building Descriptions

2.1.1 General

The WTC towers, also known as WTC 1 and WTC 2, were the primary components of the seven-building World Trade Center complex. Each of the towers encompassed 110 stories above the Plaza level and seven levels below. WTC 1 (the north tower) had a roof height of 1,368 feet, briefly earning it the title of the world's tallest building. WTC 2 (the south tower) was nearly as tall, with a roof height of 1,362 feet. WTC 1 also supported a 360-foot-tall television and radio transmission tower. Each building had a square floor plate, 207 feet 2 inches long on each side. Corners were chamfered 6 feet 11 inches. Nearly an acre of floor space was provided at each level. A rectangular service core, with overall dimensions of approximately 87 feet by 137 feet, was present at the center of each building, housing 3 exit stairways, 99 elevators, and 16 escalators. Figure 2-1 presents a schematic plan of a representative aboveground floor.

The project was developed by the Port Authority of New York and New Jersey (hereafter referred to as the Port Authority), a bi-state public agency. Original occupancy of the towers was dominated by government agencies, including substantial occupancy by the Port Authority itself. However, this occupancy evolved over the years and, by 2001, the predominant occupancy of the towers was by commercial tenants, including a number of prominent financial and insurance services firms.

Design architecture was provided by Minoru Yamasaki & Associates, and Emery Roth & Sons served as the architect of record. Skilling, Helle, Christiansen, Robertson were the project structural engineers; Jaros, Baum & Bolles were the mechanical engineers; and Joseph R. Loring & Associates were the electrical engineers. The Port Authority provided design services for site utilities, foundations, basement retaining walls, and paving. Groundbreaking for construction was on August 5, 1966. Steel construction began in August 1968. First tenant occupancy of WTC 1 was in December 1970, and occupancy of WTC 2 began in January 1972. Ribbon cutting was on April 4, 1973.

2.1.2 Structural Description

WTC 1 and WTC 2 were similar, but not identical. WTC 1 was 6 feet taller than WTC 2 and also supported a 360-foot tall transmission tower. The service core in WTC 1 was oriented east to west, and the service core in WTC 2 was oriented north to south. In addition to these basic configuration differences, the presence of each building affected the wind loads on the other, resulting in a somewhat different distribution of design wind pressures, and, therefore, a somewhat different structural design of the lateral-force-resisting system. In addition, tenant improvements over the years resulted in removal of portions of floors and placement of new private stairways between floors, in a somewhat random pattern. Figure 2-2 presents a structural framing plan representative of an upper floor in the towers.

CHAPTER 2: WTC 1 and WTC 2

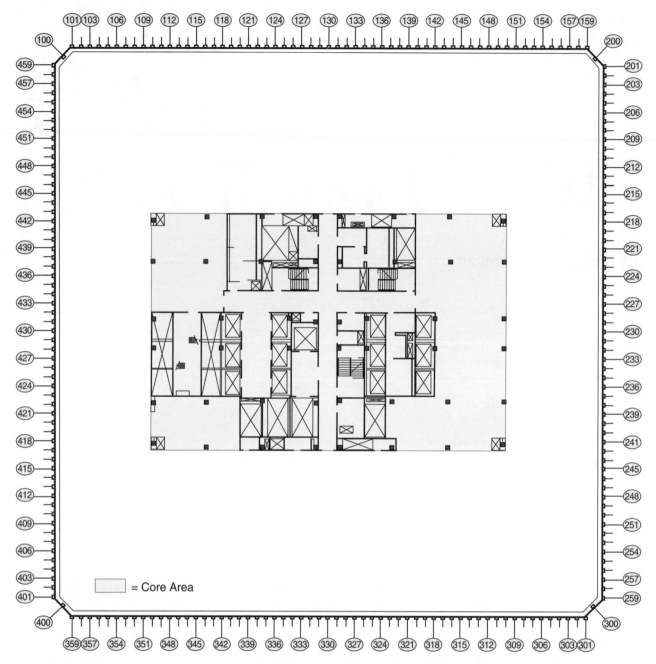

Figure 2-1 Representative floor plan (based on floor plan for 94th and 95th floors of WTC 1).

The buildings' signature architectural design feature was the vertical fenestration, the predominant element of which was a series of closely spaced built-up box columns. At typical floors, a total of 59 of these perimeter columns were present along each of the flat faces of the building. These columns were built up by welding four plates together to form an approximately 14-inch square section, spaced at 3 feet 4 inches on center. Adjacent perimeter columns were interconnected at each floor level by deep spandrel plates, typically 52 inches in depth. In alternate stories, an additional column was present at the center of each of the chamfered building corners. The resulting configuration of closely spaced columns and deep spandrels created a perforated steel bearing-wall frame system that extended continuously around the building perimeter.

Figure 2-3 presents a partial elevation of this exterior wall at typical building floors. Construction of the perimeter-wall frame made extensive use of modular shop prefabrication. In general, each exterior wall module consisted of three columns, three stories tall, interconnected by the spandrel plates, using all-welded construction. Cap plates were provided at the tops and bottoms of each column, to permit bolted connection to the modules above and below. Access holes were provided at the inside face of the columns for attaching high-strength bolted connections. Connection strength varied throughout the building, ranging from four bolts at upper stories to six bolts at lower stories. Near the building base, supplemental welds were also utilized.

Side joints of adjacent modules consisted of high-strength bolted shear connections between the spandrels at mid-span. Except at the base of the structures and at mechanical floors, horizontal splices between modules were staggered in elevation so that not more than one third of the units were spliced in any one story. Where the units were all spliced at a common level, supplemental welds were used to improve the strength of these connections. Figure 2-3 illustrates the construction of typical modules and their interconnection. At the building base, adjacent sets of three columns tapered to form a single massive column, in a fork-like formation, shown in Figure 2-4.

Twelve grades of steel, having yield strengths varying between 42 kips per square inch (ksi) and 100 ksi, were used to fabricate the perimeter column and spandrel plates as dictated by the computed gravity and wind demands. Plate thickness also varied, both vertically and around the building perimeter, to accommodate the predicted loads and minimize differential shortening of columns across the floor plate. In upper stories of the building, plate thickness in the exterior wall was generally 1/4 inch. At the base of the building, column plates as thick as 4 inches were used. Arrangement of member types (grade and thickness) was neither exactly symmetrical within a given building nor the same in the two towers.

The stiffness of the spandrel plates, created by the combined effects of the short spans and significant depth, created a structural system that was stiff both laterally and vertically. Under the effects of lateral wind loading, the buildings essentially behaved as cantilevered hollow structural tubes with perforated walls. In each building, the windward wall acted as a tension flange for the tube while the leeward wall acted as a compression flange. The side walls acted as the webs of the tube, and transferred shear between the windward and leeward walls through Vierendeel action (Figure 2-5). Vierendeel action occurs in rigid trusses that do not have diagonals. In such structures, stiffness is achieved through the flexural (bending) strength of the connected members. In the lower seven stories of the towers, where there were fewer columns (Figure 2-4), vertical diagonal braces were in place at the building cores to provide this stiffness. This structural frame was considered to constitute a tubular system.

Floor construction typically consisted of 4 inches of lightweight concrete on 1-1/2-inch, 22-gauge non-composite steel deck. In the core area, slab thickness was 5 inches. Outside the central core, the floor deck was supported by a series of composite floor trusses that spanned between the central core and exterior wall. Composite behavior with the floor slab was achieved by extending the truss diagonals above the top chord so that they would act much like shear studs, as shown in Figure 2-6. Detailing of these trusses was similar to that employed in open-web joist fabrication and, in fact, the trusses were manufactured by a joist fabricator, the LaClede Steel Corporation. However, the floor system design was not typical of open-web-joist floor systems. It was considerably more redundant and was well braced with transverse members. Trusses were placed in pairs, with a spacing of 6 feet 8 inches and spans of approximately 60 feet to the sides and 35 feet at the ends of the central core. Metal deck spanned parallel to the main trusses and was directly supported by continuous transverse bridging trusses spaced at 13 feet 4 inches and intermediate deck support angles spaced at 6 feet 8 inches from the transverse trusses. The combination of main trusses, transverse trusses, and deck support enabled the floor system to act as a grillage to distribute load to the various columns.

CHAPTER 2: *WTC 1 and WTC 2*

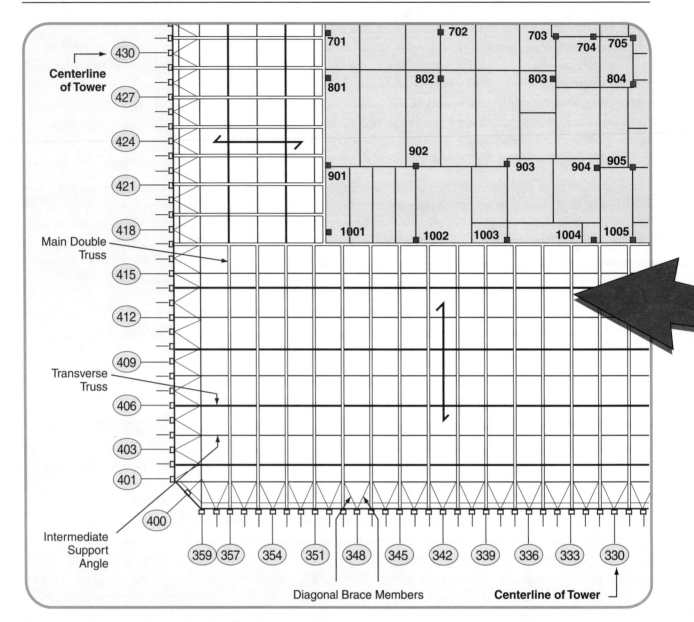

Figure 2-2   Representative structural framing plan, upper floors.

At the exterior wall, truss top chords were supported in bearing off seats extending from the spandrels at alternate columns. Welded plate connections with an estimated ultimate capacity of 90 kips (refer to Appendix B) tied the pairs of trusses to the exterior wall for out-of-plane forces. At the central core, trusses were supported on seats off a girder that ran continuously past and was supported by the core columns. Nominal out-of-plane connection was provided between the trusses and these girders. Figures 2-7 and 2-8 illustrate this construction, and Figure 2-9 shows a cross-section through typical floor framing. Floors were designed for a uniform live load of 100 pounds per square foot (psf) over any 200-square-foot area with allowable live load reductions taken over larger areas. At building corners, this amounted to a uniform live load (unreduced) of 55 psf.

At approximately 10,000 locations in each building, viscoelastic dampers extended between the lower chords of the trussses and gusset plates mounted on the exterior columns beneath the stiffened seats (Detail A in Figure 2-6). These dampers were the first application of this technology in a high-rise building, and were provided to reduce occupant perception of wind-induced building motion.

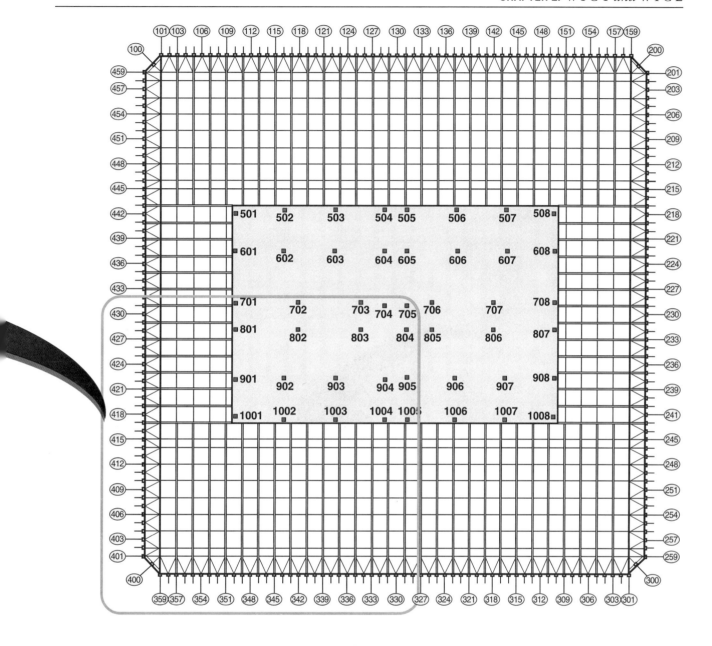

Pairs of flat bars extended diagonally from the exterior wall to the top chord of adjacent trusses. These diagonal flat bars, which were typically provided with shear studs, provided horizontal shear transfer between the floor slab and exterior wall, as well as out-of-plane bracing for perimeter columns not directly supporting floor trusses (Figure 2-2).

The core consisted of 5-inch concrete fill on metal deck supported by floor framing of rolled structural shapes, in turn supported by a combination of wide flange shape and box-section columns. Some of these columns were very large, with cross-sections measuring 14 inches wide by 36 inches deep. In upper stories, these rectangular box columns transitioned into heavy rolled wide flange shapes.

Between the 106th and 110th floors, a series of diagonal braces were placed into the building frame. These diagonal braces together with the building columns and floor framing formed a deep outrigger truss system that extended between the exterior walls and across the building core framing. A total of 10 outrigger truss lines were present in each building (Figure 2-10), 6 extending across the long direction of the core and

CHAPTER 2: *WTC 1 and WTC 2*

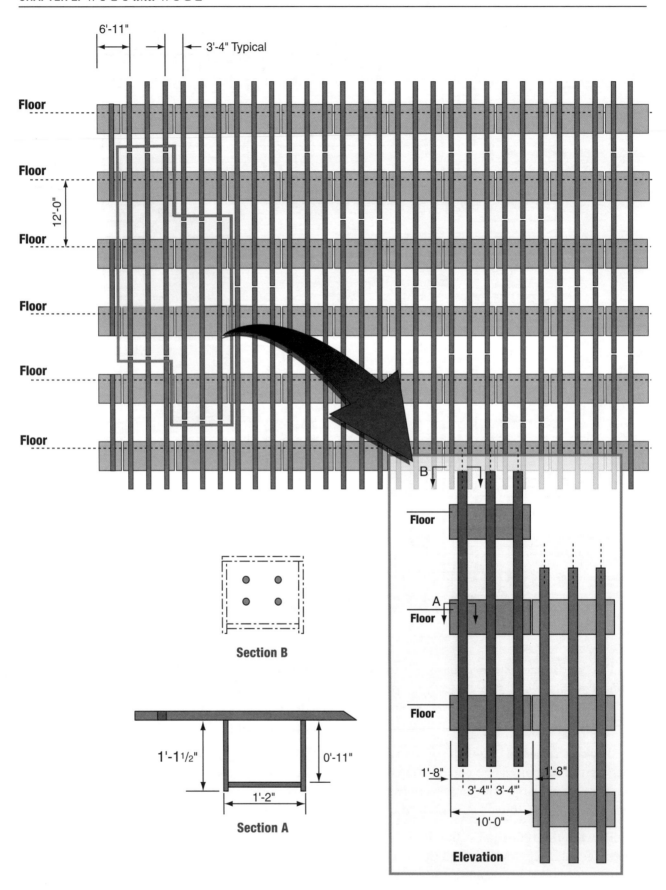

Figure 2-3   Partial elevation of exterior bearing-wall frame showing exterior wall module construction.

CHAPTER 2: *WTC 1 and WTC 2*

Figure 2-4   Base of exterior wall frame.

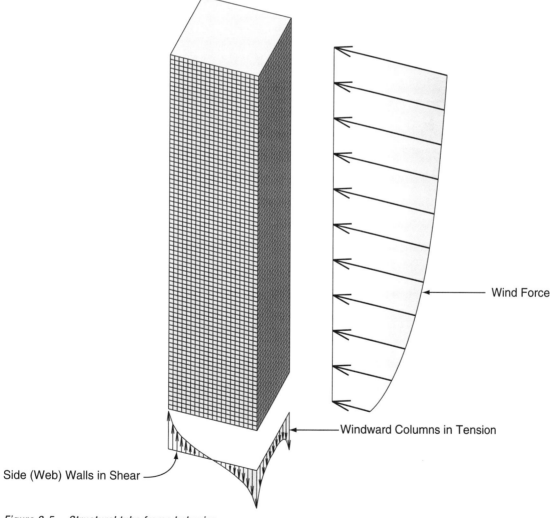

Figure 2-5   Structural tube frame behavior.

CHAPTER 2: WTC 1 and WTC 2

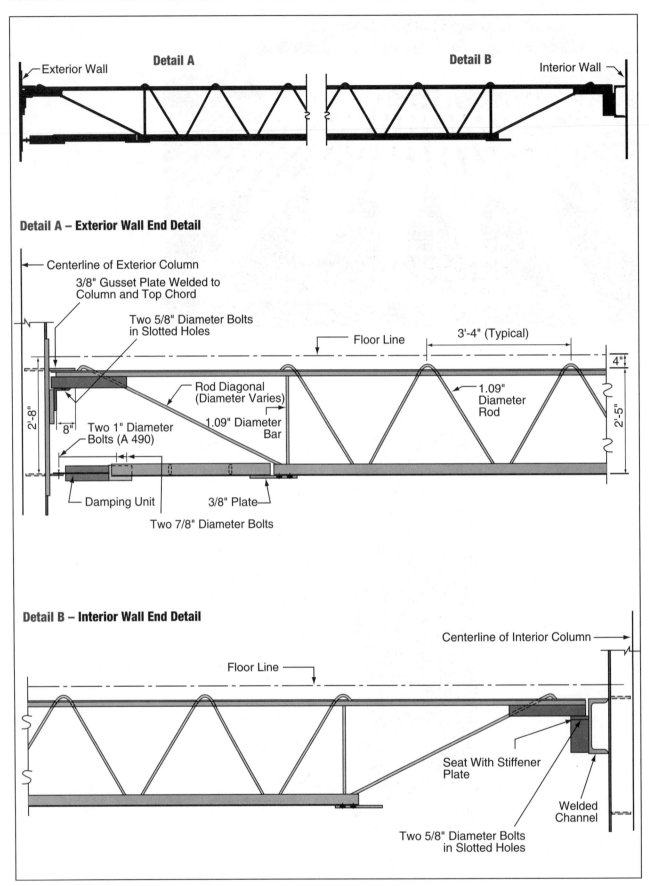

Figure 2-6   Floor truss member with details of end connections.

CHAPTER 2: *WTC 1 and WTC 2*

Figure 2-7   *Erection of prefabricated components, forming exterior wall and floor deck units.*

Figure 2-8   *Erection of floor framing during original construction.*

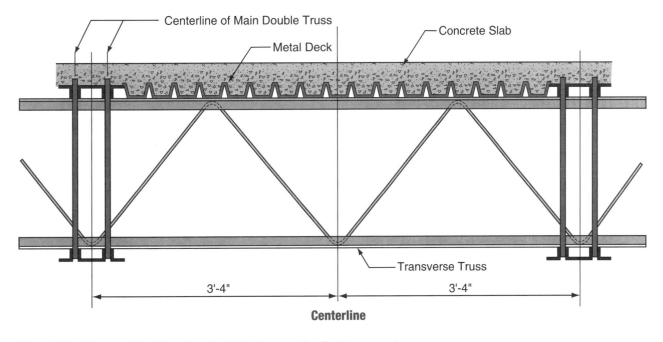

Figure 2-9   *Cross-section through main double trusses, showing transverse truss.*

FEDERAL EMERGENCY MANAGEMENT AGENCY

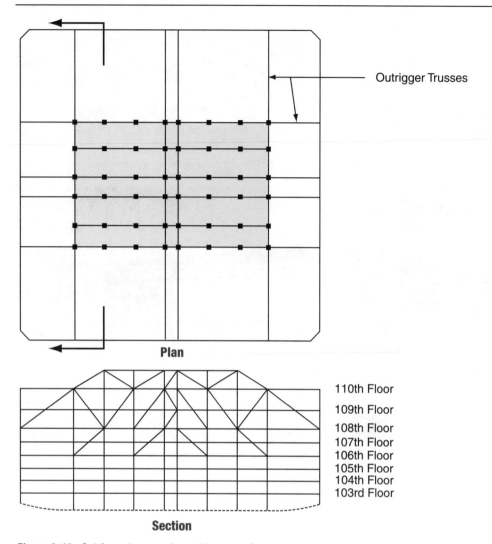

Figure 2-10 Outrigger truss system at tower roof.

4 extending across the short direction of the core. This outrigger truss system provided stiffening of the frame for wind resistance, mobilized some of the dead weight supported by the core to provide stability against wind-induced overturning, and also provided direct support for the transmission tower on WTC 1. Although WTC 2 did not have a transmission tower, the outrigger trusses in that building were also designed to support such a tower.

A deep subterranean structure was present beneath the WTC Plaza (Figure 2-11) and the two towers. The western half of this substructure, bounded by West Street to the west and by the 1/9 subway line that extends approximately between West Broadway and Greenwich Street on the east, was 70 feet deep and contained six subterranean levels. The structure housed a shopping mall and building mechanical and electrical services, and it also provided a station for the PATH subway line and parking for the complex.

Prior to construction, the site was underlain by deep deposits of fill material, informally placed over a period of several hundred years to displace the adjacent Hudson River shoreline and create additional usable land area. In order to construct this structure, the eventual perimeter walls for the subterranean structure were constructed using the slurry wall technique. After the concrete wall was cured and attained sufficient strength, excavation of the basement was initiated. As excavation proceeded downward, tieback anchors were drilled diagonally down through the wall and grouted into position in the rock deep behind the walls. These

CHAPTER 2: WTC 1 and WTC 2

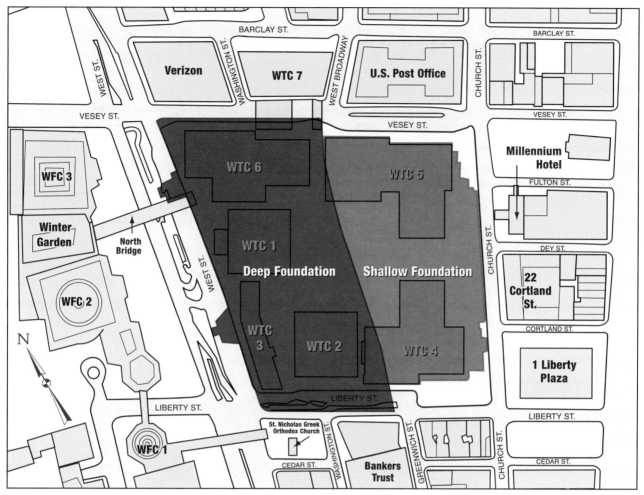

Figure 2-11 Location of subterranean structure.

anchors stabilized the wall against the soil and water pressures from the unexcavated side as the excavation continued on the inside. After the excavation was extended to the desired grade, foundations were formed and poured against the exposed bedrock, and the various subgrade levels of the structure were constructed.

Floors within the substructure were of reinforced concrete flat-slab construction, supported by structural steel columns. Many of these steel columns also provided support for the structures located above the plaza level. After the floor slabs were constructed, they were used to provide lateral support for the perimeter walls, holding back the earth pressure from the unexcavated side. The tiebacks, which had been installed as a temporary stabilizing measure, were decommissioned by cutting off their end anchorage hardware and repairing the pockets in the slurry wall where these anchors had existed.

Tower foundations beneath the substructure consisted of massive spread footings, socketed into and bearing directly on the massive bedrock. Steel grillages, consisting of layers of orthogonally placed steel beams, were used to transfer the immense column loads, in bearing, to the reinforced concrete footings.

2.1.3 Fire Protection

The fire safety of a building is provided by a system of interdependent fire protection features, including suppression systems, detection systems, notification devices, smoke management systems, and passive systems such as compartmentation and structural protection. The failure of any of these fire protection systems will impact the effectiveness of the other systems in the building.

FEDERAL EMERGENCY MANAGEMENT AGENCY

2.1.3.1 Passive Protection

In WTC 1, structural elements up to the 39th floor were originally protected from fire with a spray-applied product containing asbestos (Nicholson, et al. 1980). These asbestos-containing materials were later abated inside the building, either through encapsulation or replacement. On all other floors and throughout WTC 2, a spray-applied, asbestos-free mineral fiber material was used. Each element of the steel floor trusses was protected with spray-applied material. The specific material used was a low-density, factory-mixed product consisting of manufactured inorganic fibers, proprietary cement-type binders, and other additives in low concentrations to promote wetting, set, and dust control. Air setting, hydraulic setting, and ceramic setting binders were added in varying quantities and combinations or singly at the site, depending on the particular application and weather conditions. Finally, water was added at the nozzle of the spray gun as the material was sprayed onto the member to be protected. The average thickness of spray-applied fireproofing on the trusses was 3/4 inch. In the mid-1990s, a decision was made to upgrade the fire protection by applying additional material onto the trusses so as to increase fireproofing thickness to 1-1/2 inches. The fireproofing upgrade was applied to individual floors as they became vacant. By September 11, 2001, a total of 31 stories had been upgraded, including the entire impact zone in WTC 1 (floors 94–98), but only the 78th floor in the impact zone in WTC 2 (floors 78–84).

Spandrels and girders were specified to have sufficient protection to achieve a 3-hour rating. Except for the interior face of perimeter columns between spandrels, which were protected with a plaster material, spray-applied materials similar to those used on the floor systems were used. The thickness of protection on spandrels and girders varied, with the more massive steel column sections receiving reduced fireproofing thickness relative to the thinner elements.

The primary vertical compartmentation was provided by the floor slabs that were cast flush against the spandrel beams at the exterior wall, providing separation between floors at the building perimeter. After a fire in 1975, vertical penetrations for cabling and plumbing were sealed with fire-resistant material. At stair and elevator shafts, separation was provided by a wall system constructed of metal studs and two layers of 5/8-inch thick gypsum board on the exterior and one layer of 5/8-inch thick gypsum board on the interior. These assemblies provided a 2-hour rating. Horizontal compartmentation varied throughout the complex. Some separating walls ran from slab to slab, while others extended only up to the suspended ceiling. A report by the New York Board of Fire Underwriters (NYBFU) titled *One World Trade Center Fire, February 13, 1975* (NYBFU 1975) presents a detailed discussion of the compartmentation features of the building at that time.

2.1.3.2 Suppression

When originally constructed, the two towers were not provided with automatic fire sprinkler protection. However, such protection was installed as a retrofit circa 1990, and automatic sprinklers covered nearly 100 percent of WTC 1 and WTC 2 at the time of the September 11 attacks. In addition, each building had standpipes running through each of its three stairways. A 1.5-inch hose line and a cabinet containing two air-pressurized water (APW) extinguishers were also present at each floor in each stairway.

The primary water supply was provided by a dedicated fire yard main that looped around most of the complex. This yard main was supplied directly from the municipal water supply. Two remotely located high-pressure, multi-stage, 750-gallons per minute (gpm) electrical fire pumps took suction from the New York City municipal water supply and produced the required operating pressures for the yard main.

Each tower had three electrical fire pumps that provided additional pressure for the standpipes. One pump, located on the 7th floor, received the discharge from the yard main fire pumps and moved it up to the 41st floor, where a second 750-gpm fire pump pushed it up to a third pump on the 75th floor. Each fire pump produced sufficient pressure to supply water to the pump two stages up from it in the event that any one pump should fail.

Several 5,000-gallon storage tanks, filled from the domestic water system, provided a secondary water supply. Tanks on the 41st, 75th, and 110th floors provided water directly to a standpipe system. A tank on the 20th floor supplied water directly to the yard main. Numerous Fire Department of New York (FDNY) connections were located around the complex to allow the fire department to boost water pressure in the buildings.

2.1.3.3 Smoke Management

A zoned smoke control system was built into each building's ventilation systems and was activated upon direction of the responding FDNY Incident Commander. The system was designed to limit smoke spread from the tenant areas to the core area, thereby assisting both individuals evacuating from an area and those responding to the scene by limiting smoke spread into the core.

2.1.3.4 Fire Department Features

At the time of the 1993 World Trade Center bombing, a centralized Fire Command Center (FCC) for the two towers was present at the Concourse level. This FCC was located in the B-1 level Operations Control Center (OCC). Following the 1993 bombing, additional FCCs were installed in the lobbies of each tower.

A Radiax cable and antenna were installed in the WTC complex to facilitate the use of FDNY radios in the towers. Fire department telephones were provided in both towers on odd floors in Stairway 3, as well as on levels B-1, B-4, and B-6.

The WTC had its own fire brigade, consisting of Port Authority police officers trained in fire safety, who worked with the FDNY to investigate fire conditions and take appropriate actions. The internal fire brigade had access to fire carts located on the Concourse level and on the 44th and 78th floor sky lobbies of each tower. These fire carts were equipped with hoses, nozzles, self-contained breathing apparatus, turnout coats, forcible entry tools, resuscitators, first-aid kits, and other emergency equipment. Typically, the WTC fire brigade would collect the nearest fire cart and set up operations on the floor below the fire floor.

The WTC complex had 24 Siamese connections located at street level for use by the FDNY apparatus. Each of these Siamese connections served various portions of the complex and was identified as such.

2.1.4 Emergency Egress

Each tower was provided with three independent emergency fire exit stairways, located in the core of the building, as indicated in Figure 2-12. Two of these stairways, designated Stairway 1 and Stairway 2, were 44 inches wide and ran to the 110th floor. The third stairway, designated Stairway 3, had a width of 56 inches and ran to the 108th floor. The stairways did not run in continuous vertical shafts from the top to the bottom of the structure. Instead, the plan location of the stairways shifted at some levels, and occupants traversing the stairways were required to move from one vertical shaft to another through a transfer corridor. Both Stairways 1 and 2 had transfers at the 42nd, 48th, 76th, and 82nd levels. Stairway 1 had an additional transfer at the 26th level and Stairway 3 had a single transfer at the 76th level. After the 1993 bombing, battery-operated emergency lighting was provided in the stairways and photoluminescent paint was placed on the edge of the stair treads to facilitate emergency egress.

There were 99 elevators in each of the two towers, including 23 express elevators; however, the express elevators were not intended to be used for emergency access or egress. There were also several freight elevators servicing groups of floors in the buildings. The several elevators that served each floor were broken into two groups that operated on different power supplies.

Upon alarm activation, an automatic elevator override system commanded all elevators serving or affected by a fire area to immediately return to the ground floor, or to their sky lobby (44th and 78th floors). From there, the elevators could be operated manually by the FDNY. Although many fire departments

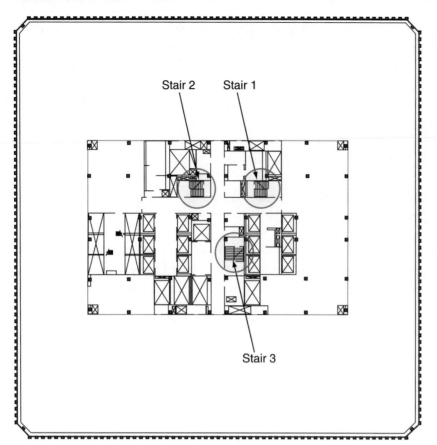

Figure 2-12
Floor plan of 94th and 95th floors of WTC 1 showing egress stairways.

routinely use elevators to provide better access in high-rise buildings, FDNY does not do this, because there have been fatalities associated with such use.

2.1.5 Emergency Power

Primary power was provided at 13.8 kilovolts (kV) through a ground level substation in WTC 7 near the Barclay Street entrance to the underground parking garage. The primary power was wired to the buildings through two separate systems. The first provided power throughout each building; the second provided power to emergency systems in the event that the primary wiring system failed.

Six 1,200-kilowatt (kW) emergency power generators located in the sixth basement (B-6) level provided a secondary power supply. These generators were checked on a routine basis to ensure that they would function properly during an emergency. This equipment provided backup power for communications equipment, elevators, emergency lighting in corridors and stairwells, and fire pumps. Telephone systems were provided with an independent battery backup system. Emergency lighting units in exit stairways, elevator lobbies, and elevator cabs were equipped with individual backup batteries.

2.1.6 Management Procedures

The Port Authority has a risk management group that coordinates fire and safety activities for their various properties. This group provided training for the WTC fire brigade, fire safety directors, and tenant fire wardens. The WTC had 25 fire safety directors who assisted in the coordination of fire safety activities in the buildings throughout the year. Six satellite communication stations, staffed by deputy fire safety directors, were spaced throughout the towers. In addition, each tenant was required to provide at least one fire warden. Tenants that occupied large areas of the building were required to provide one fire warden for every 7,500 square feet of occupied space. The fire safety directors trained the fire wardens and fire drills were held twice a year.

2.2 Building Response

WTC 1 and WTC 2 each experienced a similar, though not identical, series of loading events. In essence, each tower was subjected to three separate, but related events. The sequence of these events was the same for the two buildings, although the timing was not. In each case, the first loading event was a Boeing 767-200ER series commercial aircraft hitting the building, together with a fireball resulting from immediate rapid ignition of a portion of the fuel on board the aircraft. Boeing 767-200ER aircraft have a maximum rated takeoff weight of 395,000 pounds, a wingspan of 156 feet 1 inch, and a rated cruise speed of 530 miles per hour. The aircraft is capable of carrying up to 23,980 gallons of fuel, and it is estimated that, at the time of impact, each aircraft had approximately 10,000 gallons of unused fuel on board (compiled from Government sources).

In each case, the aircraft impacts resulted in severe structural damage, including some localized partial collapse, but did not result in the initiation of global collapse. In fact, WTC 1 remained standing for a period of approximately 1 hour and 43 minutes, following the initial impact; WTC 2 remained standing for approximately 56 minutes following impact. The second event was the simultaneous ignition and growth of fires over large floor areas on several levels of the buildings. The fires heated the structural systems and, over a period of time, resulted in additional stressing of the damaged structure, as well as sufficient additional damage and strength loss to initiate the third event, a progressive sequence of failures that culminated in total collapse of both structures.

2.2.1 WTC 1

2.2.1.1 Initial Damage From Aircraft Impact

American Airlines Flight 11 struck the north face of WTC 1 approximately between the 94th and 98th floors (Figures 2-13 and 2-14), causing massive damage to the north face of the building within the immediate area (Figure 2-15). At the central zone of impact corresponding to the airplane fuselage and engines, at least five of the prefabricated, three-column sections that formed the exterior walls were broken loose of the structure, and some were pushed inside the building envelope. Locally, floors supported by these exterior wall sections appear to have partially collapsed, losing their support along the exterior wall. Away from this central zone, in areas impacted by the outer wing structures, the exterior columns were fractured by the force of the collision. Interpretation of photographic evidence suggests that from 31 to 36 columns on the north building face were destroyed over portions of a four-story range. Partial collapse of floors in this zone appear to have occurred over a horizontal length of wall of approximately 65 feet, while floors in other portions of the building appear to have remained intact. Figure 2-16 shows the damage to the exterior columns on the impacted face of WTC 1.

In addition to this damage at the building perimeter, a significant but undefined amount of damage also occurred to framing at the central core. Interviews were conducted with persons who were present in offices on the 91st floor of the building at the north face of the structure, three floors below the approximate zone of impact. Their descriptions of the damage evident at this floor level immediately following the aircraft impact suggest relatively slight damage at the exterior wall of the building, but progressively greater damage to the south and east. They described extensive building debris in the eastern portion of the central core, preventing their access to the easternmost exit stairway. This suggests the possibility of immediate partial collapse of framing in the central core. These persons also described the presence of debris from collapsed partition walls from upper floors in stairways located further to the west, suggesting the possibility of some structural damage in the northwestern portion of the core framing as well. Figure 2-17 is a sketch made during an interview with building occupants indicating portions of the 91st floor that could not be accessed due to accumulated debris.

CHAPTER 2: *WTC 1 and WTC 2*

Figure 2-13 *Zone of aircraft impact on the north face of WTC 1.*

It is known that some debris from the aircraft traveled completely through the structure. For example, life jackets and portions of seats from the aircraft were found on the roof of the Bankers Trust building, located to the south of WTC 2. Part of the landing gear from this aircraft was found at the corner of West and Rector Streets, some five blocks south of the WTC complex (Figure 2-18). As this debris passed through the building, it doubtless caused some level of damage to the structure across the floor plate, including, potentially, interior framing, core columns, framing at the east, south, and west walls, and the floors themselves. The exact extent of this damage will likely never be known with certainty. It is evident that, despite this damage, the structure retained sufficient integrity and strength to remain globally stable for a period of approximately 1 hour and 43 minutes.

The building's structural system, composed of the exterior loadbearing frame, the gravity loadbearing frame at the central core, and the system of deep outrigger trusses in upper stories, was highly redundant. This permitted the building to limit the immediate zone of collapse to the area where several stories of exterior columns were destroyed by the initial impact and, perhaps, to portions of the central core as previously described. Following the impact, floor loads originally supported by the exterior columns in

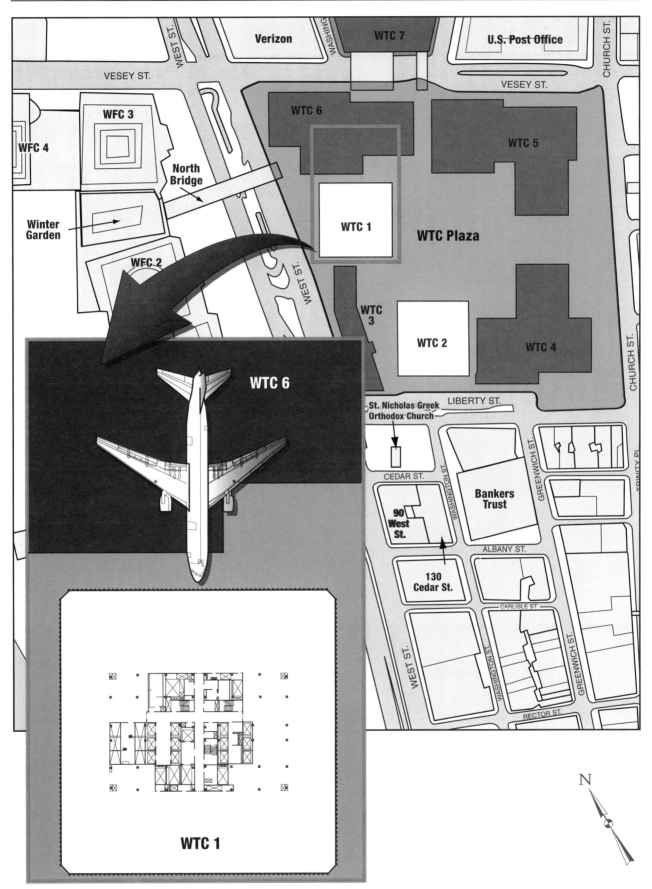

Figure 2-14 Approximate zone of impact of aircraft on the north face of WTC 1.

CHAPTER 2: *WTC 1 and WTC 2*

Figure 2-15 Impact damage to the north face of WTC 1.

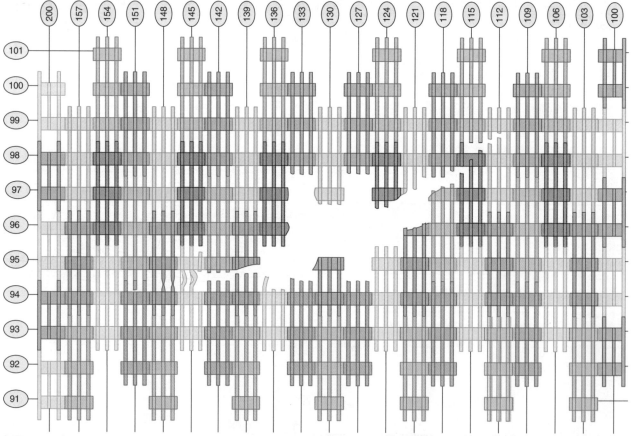

GENERAL NOTES: (1) Column damage captured from photographs and enhanced videos. (2) Damage to column lines 111-115 at level 98 is estimated.

Figure 2-16 Impact damage to exterior columns on the north face of WTC 1.

2-18 **WORLD TRADE CENTER BUILDING PERFORMANCE STUDY**

CHAPTER 2: *WTC 1 and WTC 2*

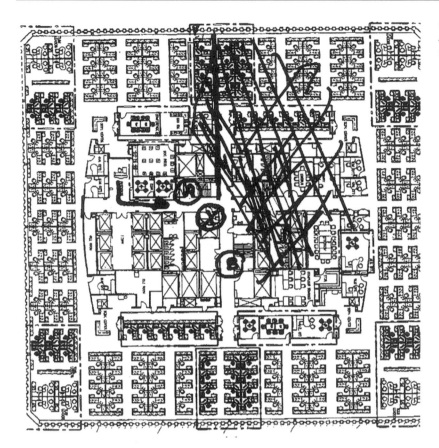

*Figure 2-17
Approximate debris location on the 91st floor of WTC 1.*

*Figure 2-18
Landing gear found at the corner of West and Rector Streets.*

compression were successfully transferred to other load paths. Most of the load supported by the failed columns is believed to have transferred to adjacent perimeter columns through Vierendeel behavior of the exterior wall frame. Preliminary structural analyses of similar damage to WTC 2 suggests that axial load demands on columns immediately adjacent to the destroyed columns may have increased by as much as a factor of 6 relative to the load state prior to aircraft impact. However, these exterior columns appear to have had substantial overstrength for gravity loads.

Neglecting the potential loss of lateral support resulting from collapsed floor slabs and any loss of strength due to elevated temperatures from fires, the most heavily loaded columns were probably near, but not over, their ultimate capacities. Columns located further from the impact zone are thought to have remained substantially below their ultimate capacities. The preliminary analyses also indicate that loss of the columns resulted in some immediate tilting of the structure toward the impact area, subjecting the remaining columns and structure to additional stresses from P-delta effects. Also, in part, exterior columns above the zone of impact were converted from compression members to hanger-type tension members, so that, in effect, a portion of the floors' weight became suspended from the outrigger trusses (Figure 2-10) and were transferred back to the interior core columns. The outrigger trusses also would have been capable of transferring some of the load carried by damaged core columns to adjacent core columns. Figure 2-19 illustrates these various secondary load paths. Section 2.2.2.2 provides a more detailed description of these analyses and findings.

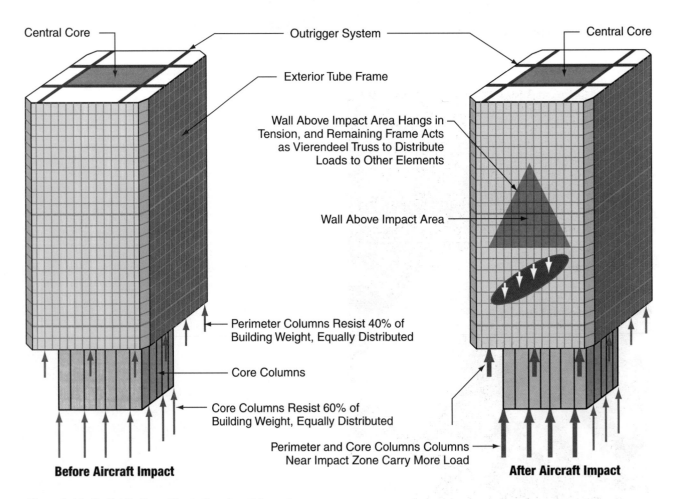

Figure 2-19 Redistribution of load after aircraft impact.

Following the aircraft impact into the building, the structure was able to successfully redistribute the building weight to the remaining elements and to maintain a stable condition. This return to a stable condition is suggested by the preliminary analyses and also evidenced by the fact that the structure remained standing for 1 hour and 43 minutes following the impact. However, the structure's global strength was severely degraded. Although the structure may have been able to remain standing in this weakened condition for an indefinite period, it had limited ability to resist additional loading and could potentially have collapsed as a result of any severe loading event, such as that produced by high winds or earthquakes. WTC 1 probably experienced some additional loading and damage due to the collapse of the adjacent WTC 2. The extent of such damage is not known but likely included broken window and façade elements along the south face. This additional damage was not sufficient to cause collapse. The first event of sufficient severity to cause collapse was the fires that followed the aircraft impact.

2.2.1.2 Fire Development

It is estimated, based on information compiled from Government sources, that each aircraft contained about 10,000 gallons of jet fuel upon impact into the buildings. A review of photographic and video records show that the aircraft fully entered the buildings prior to any visual evidence of flames at the exteriors of the buildings. This suggests that, as the aircraft crashed into and plowed across the buildings, they distributed jet fuel throughout the impact area to form a flammable "cloud." Ignition of this cloud resulted in a rapid pressure rise, expelling a fuel rich mixture from the impact area into shafts and through other openings caused by the crashes, resulting in dramatic fireballs.

Although only limited video footage is available that shows the crash of American Airlines Flight 11 into WTC 1 and the ensuing fireballs, extensive video records of the impact of United Airlines Flight 175 into WTC 2 are available. These videos show that three fireballs emanated from WTC 2 on the south, east, and west faces. The fireballs grew slowly, reaching their full size after about 2 seconds. The diameters of the fireballs were greater than 200 feet, exceeding the width of the building. Such fireballs were formed when the expelled jet fuel dispersed and flames traveled through the resulting fuel/air mixture. Experimentally based correlations for similar fireballs (Zalosh 1995) were used to estimate the amount of fuel consumed. The precise size of the fireballs and their exact shapes are not well defined; therefore, there is some uncertainty associated with estimates of the amount of fuel consumed by these effects. Calculations indicate that between 1,000 and 3,000 gallons of jet fuel were likely consumed in this manner. Barring additional information, it is reasonable to assume that an approximately similar amount of jet fuel was consumed by fireballs as the aircraft struck WTC 1.

Although dramatic, these fireballs did not explode or generate a shock wave. If an explosion or detonation had occurred, the expansion of the burning gasses would have taken place in microseconds, not the 2 seconds observed. Therefore, although there were some overpressures, it is unlikely that the fireballs, being external to the buildings, would have resulted in significant structural damage. It is not known whether the windows that were broken shortly after impact were broken by these external overpressures, overpressures internal to the building, the heat of the fire, or flying debris.

The first arriving firefighters observed that the windows of WTC 1 were broken out at the Concourse level. This breakage was most likely caused by overpressure in the elevator shafts. Damage to the walls of the elevator shafts was also observed as low as the 23rd floor, presumably as a result of the overpressures developed by the burning of the vapor cloud on the impact floors.

If one assumes that approximately 3,000 gallons of fuel were consumed in the initial fireballs, then the remainder either escaped the impact floors in the manners described above or was consumed by the fire on the impact floors. If half flowed away, then approximately 4,000 gallons remained on the impact floors to be consumed in the fires that followed. The jet fuel in the aerosol would have burned out as fast as the flame could spread through it, igniting almost every combustible on the floors involved. Fuel that fell to the floor and did not flow out of the building would have burned as a pool or spill fire at the point where it came to rest.

The time to consume the jet fuel can be reasonably computed. At the upper bound, if one assumes that all 10,000 gallons of fuel were evenly spread across a single building floor, it would form a pool that would be consumed by fire in less than 5 minutes (SFPE 1995) provided sufficient air for combustion was available. In reality, the jet fuel would have been distributed over multiple floors, and some would have been transported to other locations. Some would have been absorbed by carpeting or other furnishings, consumed in the flash fire in the aerosol, expelled and consumed externally in the fireballs, or flowed away from the fire floors. Accounting for these factors, it is believed that almost all of the jet fuel that remained on the impact floors was consumed in the first few minutes of the fire.

As the jet fuel burned, the resulting heat ignited office contents throughout a major portion of several of the impact floors, as well as combustible material within the aircraft itself.

A limited amount of physical evidence about the fires is available in the form of videos and still photographs of the buildings and the smoke plume generated soon after the initial attack. Estimates of the buoyant energy in the plume were obtained by plotting the rise of the smoke plume, which is governed by buoyancy in the vertical direction and by the wind in the horizontal direction. Using the Computational Fluid Dynamics (CFD) fire model, Fire Dynamics Simulator Ver. 1 (FDS1), fire scientists at the National Institute of Standards and Technology (NIST) (Rehm, et al. 2002) were able to mathematically approximate the size of fires required to produce such a smoke plume. As input to this model, an estimate of the openings available to provide ventilation for the fires was obtained from an examination of photographs taken of the damaged tower. Meteorological data on wind velocity and atmospheric temperatures were provided by the National Oceanic and Atmospheric Administration (NOAA) based on reports from the Aircraft Communications Addressing and Reporting System (ACARS). The information used weather monitoring instruments onboard three aircraft that departed from LaGuardia and Newark airports between 7:15 a.m. and 9:00 a.m. on September 11, 2001. The wind speed at heights equal to the upper stories of the towers was in the range of 10–20 mph. The outside temperatures over the height of the building were 20–21 °C (68–70 °F).

The modeling suggests a peak total rate of fire energy output on the order of 3–5 trillion Btu/hr, around 1–1.5 gigawatts (GW), for each of the two towers. From one third to one half of this energy flowed out of the structures. This vented energy was the force that drove the external smoke plume. The vented energy and accompanying smoke from both towers combined into a single plume. The energy output from each of the two buildings is similar to the power output of a commercial power generating station. The modeling also suggests ceiling gas temperatures of 1,000 °C (1,800 °F), with an estimated confidence of plus or minus 100 °C (200 °F) or about 900–1,100 °C (1,600–2,000 °F). A major portion of the uncertainty in these estimates is due to the scarcity of data regarding the initial conditions within the building and how the aircraft impact changed the geometry and fuel loading. Temperatures may have been as high as 900–1,100 °C (1,700–2,000 °F) in some areas and 400–800 °C (800–1,500 °F) in others.

The viability of a 3–5 trillion Btu/hr (1–1.15 GW) fire depends on the fuel and air supply. The surface area of office contents needed to support such a fire ranges from about 30,000–50,000 square feet, depending on the composition and final arrangement of the contents and the fuel loading present. Given the typical occupied area of a floor as approximately 30,000 square feet, it can be seen that simultaneous fire involvement of an area equal to 1–2 entire floors can produce such a fire. Fuel loads are typically described in terms of the equivalent weight of wood. Fuel loads in office-type occupancies typically range from about 4–12 psf, with the mean slightly less than 8 psf (Culver 1977). File rooms, libraries, and similar concentrations of paper materials have significantly higher concentrations of fuel. At the burning rate necessary to yield these fires, a fuel load of about 5 psf would be required to provide sufficient fuel to maintain the fire at full force for an hour, and twice that quantity to maintain it for 2 hours. The air needed to support combustion would be on the order of 600,000–1,000,000 cubic feet per minute.

Air supply to support the fires was primarily provided by openings in the exterior walls that were created by the aircraft impacts and fireballs, as well as by additional window breakage from the ensuing heat of the fires. Table 2.1 lists the estimated exterior wall openings used in these calculations. Although the table shows the openings on a floor-by-floor basis, several of the openings, particularly in the area of impact, actually spanned several floors (see Figure 2-17).

Sometimes, interior shafts in burning high-rise buildings also deliver significant quantities of air to a fire, through a phenomenon known as "stack effect," which is created when differences between the ambient exterior air temperatures and the air temperatures inside the building result in differential air pressures, drawing air up through the shafts to the fire area. Because outside and inside temperatures appear to have been virtually the same on September 11, this stack effect was not expected to be strong in this case.

Based on photographic evidence, the fire burned as a distributed collection of large but separate fires with significant temperature variations from space to space, depending on the type and arrangement of combustible material present and the available air for combustion in each particular space. Consequently, the temperature and related incident heat flux to the structural elements varied with both time and location. This information is not currently available, but could be modeled with advanced CFD fire models.

Damage caused by the aircraft impacts is believed to have disrupted the sprinkler and fire standpipe systems, preventing effective operation of either the manual or automatic suppression systems. Even if these systems had not been compromised by the impacts, they would likely have been ineffective. It is believed that the initial flash fires of jet fuel would have opened so many sprinkler heads that the systems would have quickly depressurized and been unable to effectively deliver water to the large area of fire involvement. Further, the initial spread of fires was so extensive as to make occupant use of small hose streams ineffective.

Table 2.1 Estimated Openings in Exterior Walls of WTC 1

Floor	North Wall ft²	North Wall (m²)	South Wall ft²	South Wall (m²)	East Wall ft²	East Wall (m²)	West Wall ft²	West Wall (m²)	Total Area ft²	Total Area (m²)
92	743	(69)	0	(0)	1,572	(146)	0	(0)	2,314	(215)
93	958	(89)	0	(0)	1,356	(126)	0	(0)	2,314	(215)
94	592	(55)	54	(5)	1,163	(108)	0	(0)	1,808	(168)
95	1,055	(98)	54	(5)	0	(0)	420	(39)	1,528	(142)
96	797	(74)	151	(14)	0	(0)	1,518	(141)	2,465	(229)
97	926	(86)	151	(14)	0	(0)	1,798	(167)	2,874	(267)
98	1,335	(124)	0	(0)	0	(0)	0	(0)	1,335	(124)
TOTAL	6,405	(595)	409	(38)	4,090	(380)	3,735	(347)	14,639	(1,360)

NOTE: Differences in totals are due to rounding in the conversion of square meters to square feet.

2.2.1.3 Evacuation

Some occupants of WTC 1 and WTC 2 began to voluntarily evacuate the buildings soon after the first aircraft struck WTC 1. Full evacuation of all occupants below the impact floors in WTC 1 was ordered soon after the second plane hit the south tower (Smith 2002). As indicated by Cauchon (2001a), the overall evacuation of the towers was as much of a success as thought possible, given the overall incident. Cauchon indicates that, between both towers, 99 percent of the people below the floors of impact survived (2001a) and by the time WTC 2 collapsed, the stairways in WTC 1 were observed to be virtually clear of building occupants (Smith 2002). In part this was possible because conditions in the stairways below the impact levels largely remained tenable. However, this may also be a result of physical changes and training programs put into place following the 1993 WTC bombing. Important modifications to building egress made following the 1993 WTC bombing included the placement of photo-luminescent paint on the egress paths to assist in wayfinding (particularly at the stair transfer corridors) and provision of emergency lighting for the stairways. In addition, an evacuation training program was instituted (Masetti 2001).

Shortly before the times of collapse, the stairways were reported as being relatively clear, indicating that occupants who were physically capable and had access to egress routes were able to evacuate from the buildings (Mayblum 2001). People within and above the impact area could not evacuate, simply because the stairways in the impact area had been destroyed.

Some survivors reported that, at about the same time that WTC 2 collapsed, lighting in the stairways of WTC 1 was lost (Mayblum 2001). Also, there were several accounts of water flowing down the stairways and of stairwells becoming slippery beginning at the 10th floor (Labriola 2001).

Anecdotes indicate altruistic behavior was commonly displayed. Some mobility-impaired occupants were carried down many flights of stairs by other occupants. There were also reports of people frequently stepping aside and temporarily stopping their evacuation to let burned and badly injured occupants pass by (Dateline NBC 2001, Hearst 2001). Occupants evacuating from the 91st floor noted that, as they descended to lower levels of the building, traffic was considerably impaired and formed into a slowly moving single-file progression, as evacuees worked their way around firefighters and other emergency responders, who were working their way up the stairways or who were resting from the exertion of the climb (Shark and McIntyre 2001).

2.2.1.4 Structural Response to Fire Loading

As previously indicated, the impact of the aircraft into WTC 1 substantially degraded the strength of the structure to withstand additional loading and also made the building more susceptible to fire-induced failure. Among the most significant factors:

- The force of the impact and the resulting debris field and fireballs probably compromised spray-applied fire protection of some steel members in the immediate area of impact. The exact extent of this damage will probably never be known, but this likely resulted in greater susceptibility of the structure to fire-related failure.

- Some of the columns were under elevated states of stress following the impact, due to the transfer of load from the destroyed and damaged elements.

- Some portions of floor framing directly beneath the partially collapsed areas were carrying substantial additional weight from the resulting debris and, in some cases, were likely carrying greater loads than they were designed to resist.

As fire spread and raised the temperature of structural members, the structure was further stressed and weakened, until it eventually was unable to support its immense weight. Although the specific chain of events that led to the eventual collapse will probably never be identified, the following effects of fire on structures may each have contributed to the collapse in some way. Appendix A presents a more detailed discussion of the structural effects of fire.

- As floor framing and supported slabs above and in a fire area are heated, they expand. As a structure expands, it can develop additional, potentially large, stresses in some elements. If the resulting stress state exceeds the capacity of some members or their connections, this can initiate a series of failures (Figure 2-20).

- As the temperature of floor slabs and support framing increases, these elements can lose rigidity and sag into catenary action. As catenary action progresses, horizontal framing elements and floor slabs become tensile elements, which can cause failure of end connections (Figure 2-21) and allow supported floors to collapse onto the floors below. The presence of large amounts of debris on some floors of WTC 1 would have made them even more susceptible to this behavior. In addition to overloading the floors below, and potentially resulting in a pancake-type collapse of successive floors, local floor collapse would also immediately increase the laterally unsupported length of columns, permitting buckling to begin. As indicated in Appendix B, the propensity of exterior columns to buckle would have been governed by the relatively weak bolted column splices between the vertically stacked prefabricated exterior wall units. This effect would be even more likely to occur in a fire that involves several adjacent floor levels simultaneously, because the columns could effectively lose lateral support over several stories (Figure 2-22).

- As the temperature of column steel increases, the yield strength and modulus of elasticity degrade and the critical buckling strength of the columns will decrease, potentially initiating buckling, even if lateral support is maintained. This effect is most likely to have been significant in the failure of the interior core columns.

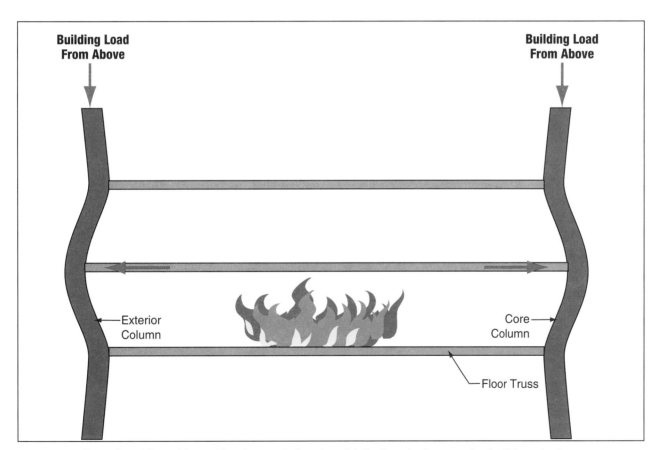

Figure 2-20 Expansion of floor slabs and framing results in outward deflection of columns and potential overload.

CHAPTER 2: *WTC 1 and WTC 2*

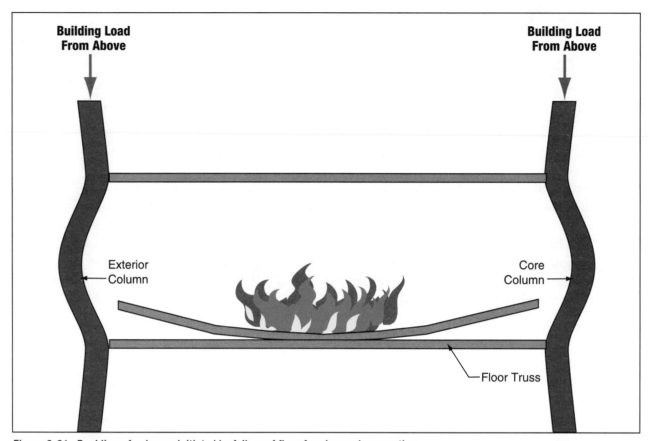

Figure 2-21 Buckling of columns initiated by failure of floor framing and connections.

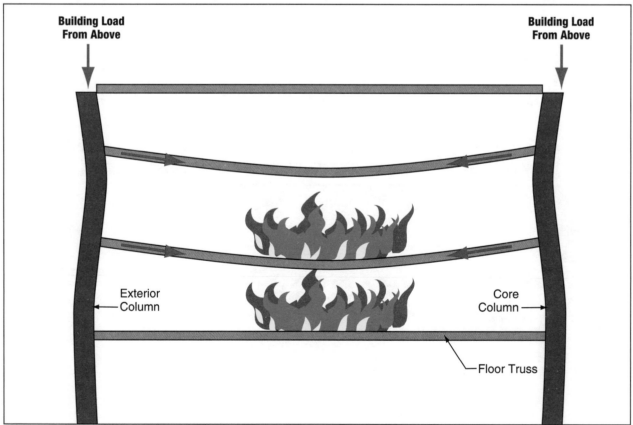

Figure 2-22 Catenary action of floor framing on several floors initiates column buckling failures.

2.2.1.5 Progression of Collapse

Construction of WTC 1 resulted in the storage of more than 4×10^{11} joules of potential energy over the 1,368-foot height of the structure. Of this, approximately 8×10^9 joules of potential energy were stored in the upper part of the structure, above the impact floors, relative to the lowest point of impact. Once collapse initiated, much of this potential energy was rapidly converted into kinetic energy. As the large mass of the collapsing floors above accelerated and impacted on the floors below, it caused an immediate progressive series of floor failures, punching each in turn onto the floor below, accelerating as the sequence progressed. As the floors collapsed, this left tall freestanding portions of the exterior wall and possibly central core columns. As the unsupported height of these freestanding exterior wall elements increased, they buckled at the bolted column splice connections, and also collapsed. Perimeter walls of the building seem to have peeled off and fallen directly away from the building face, while portions of the core fell in a somewhat random manner. The perimeter walls broke apart at the bolted connections, allowing individual prefabricated units that formed the wall or, in some cases, large assemblies of these units to fall to the street and onto neighboring buildings below.

Review of videotape recordings of the collapse taken from various angles indicates that the transmission tower on top of the structure began to move downward and laterally slightly before movement was evident at the exterior wall. This suggests that collapse began with one or more failures in the central core area of the building. This is consistent with the observations of debris patterns from the 91st floor, previously discussed. This is also supported by preliminary evaluation of the load carrying capacity of these columns, discussed in more detail in Section 2.2.2.2. The core columns were not designed to resist wind loads and, therefore, had less reserve capacity than perimeter columns. As some exterior and core columns were damaged by the aircraft impact, the outrigger trusses at the top of the building shifted additional loads to the remaining core columns, further eroding the available factor of safety. This would have been particularly significant in the upper portion of the damaged building. In this region, the original design load for the core columns was less than at lower floors, and the column sections were relatively light. The increased stresses caused by the aircraft impact could easily have brought several of these columns close to their ultimate capacity, so that relatively little additional effects due to fire would have been required to initiate the collapse. Once movement began, the entire portion of the building above the area of impact fell in a unit, pushing a cushion of air below it. As this cushion of air pushed through the impact area, the fires were fed by new oxygen and pushed outward, creating the illusion of a secondary explosion.

Although the building appeared to collapse within its own footprint, a review of aerial photographs of the site following the collapse, as well as damage to adjacent structures, suggests that debris impacted the Marriott Hotel (WTC 3), the Customs House (WTC 6), the Morgan Stanley building (WTC 5), WTC 7, and the American Express and Winter Garden buildings located across West Street (Figure 2-23). The debris field extended as far as 400–500 feet from the tower base.

2.2.2 WTC 2

2.2.2.1 Initial Damage From Aircraft Impact

United Airlines Flight 175 struck the south face of WTC 2 approximately between the 78th and 84th floors. The zone of impact extended from near the southeast corner of the building across much of the building face (Figures 2-24 and 2-25). The aircraft caused massive damage to the south face of the building in the zone of impact (Figures 2-26 and 2-27). At the central zone of impact corresponding to the airplane fuselage and engines, six of the prefabricated, three-column sections that formed the exterior walls were broken loose from the structure, with some of the elements apparently pushed inside the building envelope. Locally, as was the case in WTC 1, floors supported by these exterior wall sections appear to have partially collapsed. Away from this central zone, in the areas impacted by the outer wing structures, the exterior steel columns

CHAPTER 2: *WTC 1 and WTC 2*

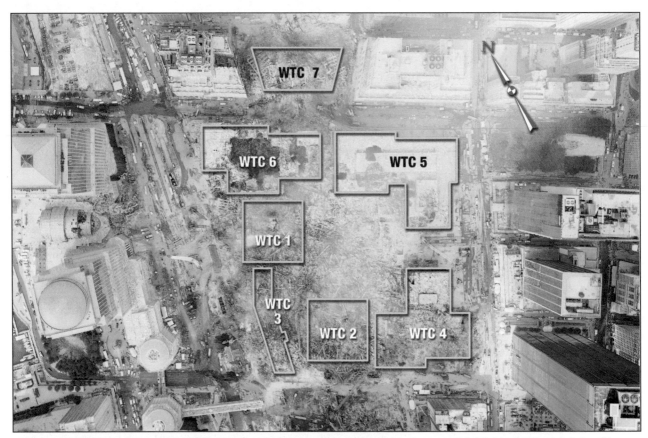

Figure 2-23 Aerial photograph of the WTC site after September 11 attack showing adjacent buildings damaged by debris from the collapse of WTC 1.

Figure 2-24 Southeast corner of WTC 2 shortly after aircraft impact.

CHAPTER 2: *WTC 1 and WTC 2*

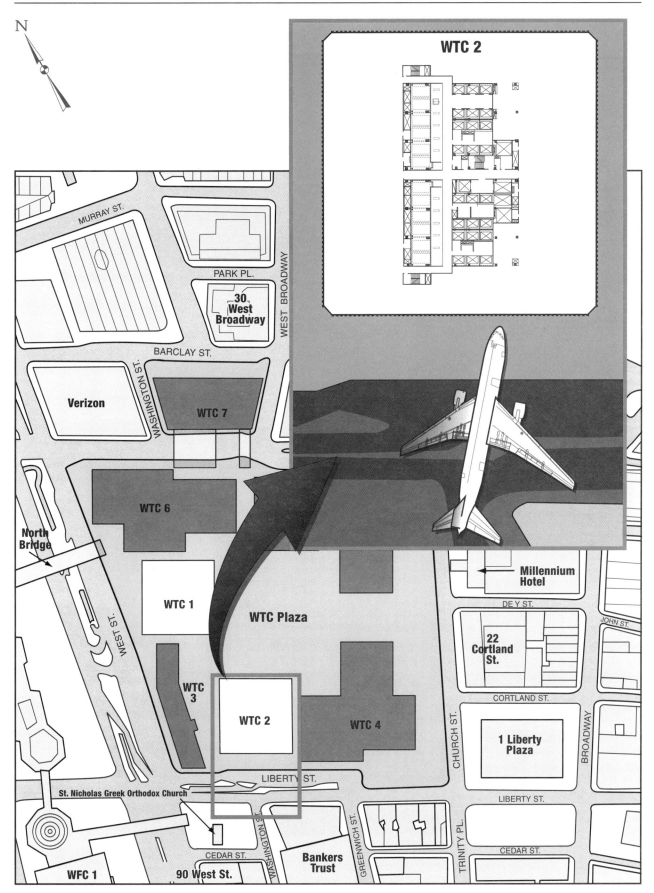

Figure 2-25   Approximate zone of impact of aircraft on the south face of WTC 2.

FEDERAL EMERGENCY MANAGEMENT AGENCY

CHAPTER 2: *WTC 1 and WTC 2*

Figure 2-26 Impact damage to the south and east faces of WTC 2.

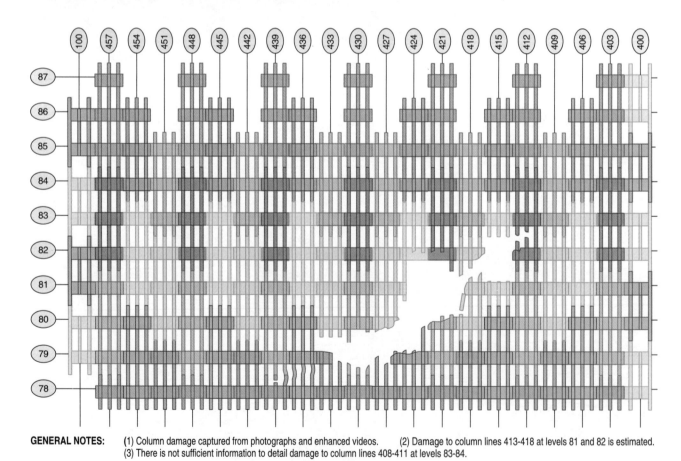

GENERAL NOTES: (1) Column damage captured from photographs and enhanced videos. (2) Damage to column lines 413-418 at levels 81 and 82 is estimated. (3) There is not sufficient information to detail damage to column lines 408-411 at levels 83-84.

Figure 2-27 Impact damage to exterior columns on the south face of WTC 2.

2-30 **WORLD TRADE CENTER BUILDING PERFORMANCE STUDY**

were fractured by the impact. Photographic evidence suggests that from 27 to 32 columns along the south building face were destroyed over a five-story range. Partial collapse of floors in this zone appears to have occurred over a horizontal length of approximately 70 feet, while floors in other portions of the building appeared to remain intact. It is probable that the columns in the southeast corner of the core also experienced some damage because they would have been in the direct travel path of the fuselage and port engine (Figure 2-25).

It is known that debris from the aircraft traveled completely through the structure. For example, a landing gear from the aircraft that impacted WTC 2 was found to have crashed through the roof of a building located six blocks to the north, and one of the jet engines was found at the corner of Murray and Church Streets. The extent to which debris scattered throughout the impact floors is also evidenced by photographs of the fireballs that occurred as the aircraft struck the building (Figure 2-28). Figure 2-29 shows a portion of the fuselage of the aircraft, lying on the roof of WTC 5.

As described for WTC 1, this debris doubtless caused some level of damage to the structure across the floor plates, including interior framing; core columns at the southeast corner of the core; framing at the north, east, and west walls; and the floor plates themselves. Figure 2-30, showing the eastern side of the north face of the WTC 2 partially hidden behind WTC 1, suggests that damage to the exterior walls was not severe except at the zone of impact. The exact extent of this damage will likely never be known with certainty. It is evident that the structure retained sufficient integrity and strength to remain globally stable for a period of approximately 56 minutes.

There are some important differences between the impact of the aircraft into WTC 2 and the impact into WTC 1. First, United Airlines Flight 175 was flying much faster, with an estimated speed of 590 mph, while American Airlines Flight 11 was flying at approximately 470 mph. The additional speed would have

Figure 2-28 Conflagration and debris exiting the north wall of WTC 2, behind WTC 1.

CHAPTER 2: WTC 1 and WTC 2

Figure 2-29 A portion of the fuselage of United Airlines Flight 175 on the roof of WTC 5.

given the aircraft a greater ability to destroy portions of the structure. The zone of aircraft impact was skewed toward the southeast corner of WTC 2, while the zone of impact on WTC 1 was approximately centered on the building's north face. The orientation of the core in WTC 2 was such that the aircraft debris would only have to travel 35 feet across the floor before it began to impact and damage elements of the core structure. Finally, the zone of impact in WTC 2 was nearly 20 stories lower than that in WTC 1, so columns in this area were carrying substantially larger loads. It is possible, therefore, that structural damage to WTC 2 was more severe than that to WTC 1, partly explaining why WTC 2 collapsed more quickly than WTC 1.

2.2.2.2 Preliminary Structural Analysis

An approximate linear structural analysis of WTC-2 was performed using SAP-2000 software (CSI 2000) to provide an understanding of the likely stress state in the building following the aircraft impact. The upper 55 stories of the building's exterior-wall frame were explicitly modeled using beam and column elements. This encompassed the entire structure above the zone of impact and about 20 stories below. The lower 55 stories of the exterior were modeled as a "boundary condition" consisting of a perimeter super-beam that was 52 inches deep and about 50 inches wide, supported on a series of springs. A base spring was provided at each column location to represent the axial stiffness of the columns from the 55th floor down to grade. The outrigger trusses at the top of the building were explicitly modeled, using truss-type elements. The interior core columns were modeled as spring elements.

An initial analysis of the building was conducted to simulate the pre-impact condition. In addition to the weight of the floor itself (approximately 54 psf at the building edges and 58 psf at the building sides), a uniform floor loading of 12 psf was assumed for partitions and an additional 20 psf was conservatively assumed to represent furnishings and contents. At the 80th floor level, exterior columns were found to be

Figure 2-30
North face of WTC 2 opposite the zone of impact on the south face, behind WTC 1.

approximately uniformly loaded with an average utilization ratio (ratio of actual applied stress to ultimate stress) of under 20 percent. This low utilization ratio is due in part to the unusually close spacing of the columns in this building, which resulted in a very small tributary area for each column. It reflects the fact that wind and deflection considerations were dominant factors in the design. Core columns were more heavily loaded with average calculated utilization ratios of 60 percent, which would be anticipated for these columns, which were designed to resist only gravity loads.

A second analysis was conducted to estimate the demands on columns immediately following aircraft impact and before fire effects occurred. Exterior columns were removed from the model to match the damage pattern illustrated in Figure 2-27. Although some core columns were probably damaged by the aircraft impact, the exact extent of this damage is not known and therefore was not considered in the model. As a result, this analysis is thought to underestimate the true stress state in the columns immediately after impact. The analysis indicates that most of the loads initially carried by the damaged exterior columns were transferred by Vierendeel truss action to the remaining exterior columns immediately adjacent to the impact area. If the floors at this level are assumed to remain intact and capable of providing lateral support to the columns, this raised the utilization ratio for the most heavily loaded column immediately adjacent to the damage area to approximately a value of 1.0. At a value of 1.0, columns would lose stiffness and shift load to adjacent columns. Based on this analysis, it appears that the structure had significant remaining margin against collapse. However, this analysis does not consider damage to the building core, which was likely significant. Columns located further from the damage area are less severely impacted, and columns located only 20 feet away from the damaged area experience almost no increase in demand at all. These data are plotted in Figure 2-31.

The columns immediately above the damage area are predicted to act as tension members, transferring approximately 10 percent of the load initially carried by the damaged columns upward to the outrigger trusses, which, in turn, transfer this load back to the core columns. Not considering any damage to the core columns, utilization ratios on these elements are predicted to increase by about 20 percent at the 80th floor level. In upper stories, where the core columns were more lightly loaded, the increase in utilization ratio is substantially larger and may have approached a value of 1.0. These conditions would have been made more severe by damage to one or more core columns.

2.2.2.3 Fire Development

Following the impact, fires spread throughout WTC 2, similar to the manner previously described for WTC 1. Extensive videotape of the fires' development through the building was recorded from various exterior vantage points. This videotape suggests that, in the minutes immediately preceding the collapse, the most intensive fires occurred along the north face of the building, near the 80th floor level. Just prior to the collapse, a stream of molten material—possibly aluminum from the airliner—was seen streaming out of a window opening at the northeast corner at approximately this level. This is of particular interest because, although the building collapse appears to have initiated at this floor level, the initiation seems to have occurred at the southeast rather than the northeast corner.

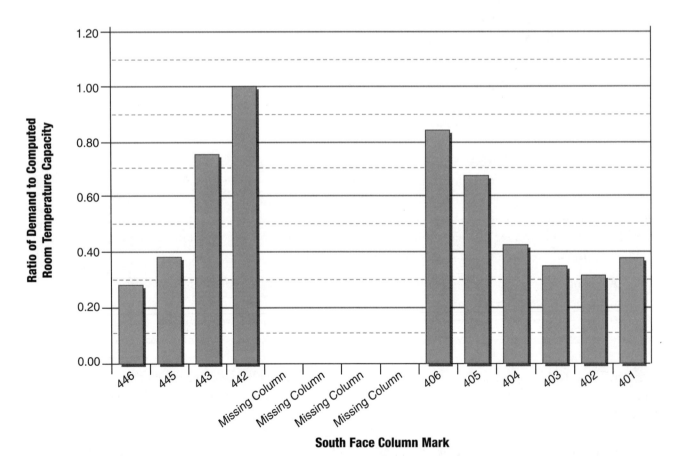

Figure 2-31 Plot of column utilization ratio at the 80th floor of WTC 2, viewed looking outward. (Conservatively assumes columns 407–411 and 440–441 to be missing.)

2.2.2.4 Evacuation

Although less time was available for evacuation of WTC 2 than for WTC 1, and the aircraft hit the building some 16 floors lower than in WTC 1, fewer casualties occurred within this building. The reduced number of casualties to building occupants in WTC 2 may be attributed to the movement of some of the building occupants immediately after the aircraft impact into WTC 1 and before the second aircraft struck WTC 2. Several survivors from WTC 2 stated that, following the impact of the aircraft into WTC 1, a message was broadcast over the loudspeaker system indicating that WTC 2 was secure and that occupants should return to their offices (Scripps 2001, BBC News 2001). Many of these survivors did not heed the announcement and continued to exit the building, using the elevators. Survivors also related reports of individuals who listened to the message, returned to their floors, and did not make it out after the second aircraft impacted WTC 2. Some survivors related that a small number of people traveled to the roof under the assumption that a helicopter rescue was possible (Cauchon 2001b).

2.2.2.5 Initiation of Collapse

The same types of structural behaviors and failure mechanisms previously discussed are equally likely to have occurred in WTC 2, resulting in the initiation of progressive collapse, approximately 56 minutes after the aircraft impact. Review of video footage of the WTC 2 collapse suggests that it probably initiated with a partial collapse of the floor in the southeast corner of the building at approximately the 80th level. This appears to have been followed rapidly by collapse of the entire floor level along the east side, as evidenced by a line of dust blowing out of the side of the building. As this floor collapse occurred, columns along the east face of the building appear to buckle in the region of the collapsed floor, beginning at the south side and progressing to the north, causing the top of the building to rotate toward the east and south and to begin to collapse downward (Figure 2-32). It should be noted that failure of core columns in the southeast corner of the building could have preceded and triggered these events.

2.2.2.6 Progression of Collapse

As in WTC 1, a very large quantity of potential energy was stored in the building, during its construction. Once collapse initiated, much of this energy was rapidly released and converted into kinetic energy, in the form of the rapidly accelerating mass of the structure above the aircraft impact zone. The impact of this rapidly moving mass on the lower structure caused a wide range of structural failures in the floors directly at and below the aircraft impact zone, in turn causing failure of these floors. As additional floor plates failed, the mass associated with each of these floors joined that of the tower above the impact area, increasing the destructive energy on the floors immediately below. This initiated a chain of progressive failures that resulted in the total collapse of the building.

A review of aerial photographs of the site, following the collapse, as well as identification of pieces of structural steel from WTC 2, strongly suggests that while the top portion of the tower fell to the south and east, striking Liberty Street and the Bankers Trust building, the lower portion of the tower fell to the north and west, striking the Marriott Hotel (WTC 3). Again, the debris pattern spread laterally as far as approximately 400–500 feet from the base of the structure.

2.2.3 Substructure

As first WTC 2, then WTC 1 collapsed, nearly 600,000 tons of debris fell onto the Plaza level, punching large holes through the Plaza and the six levels of substructure below, and partially filling the substructure with debris. This damage severely compromised the ability of the slabs to provide lateral bracing of the substructure walls against the induced lateral earth pressures from the unexcavated side. This condition was most severe at the southern side of the substructure, adjacent to WTC 2 and WTC 3. In this region, debris from the collapsed WTC 2 punched through several levels of substructure slab, but

Figure 2-32 The top portion of WTC 2 falls to the east, then south, as viewed from the northeast.

did not completely fill the void left behind, leaving the south wall of the substructure in an unbraced condition over a portion of its length.

In early October, large cracks were observed along Liberty Street, indicating that the south wall had started to move into the failed area under the influence of the lateral earth pressures. Mueser-Rutledge Engineers were retained to review the situation and make suitable recommendations. As a temporary measure, sand fill was backfilled against the inside face of the south wall to counterbalance earth pressures on the unexcavated side. Following temporary stabilization of the wall, tiebacks were reinstalled through the wall in a manner similar to that used to stabilize the excavation during the original construction of the development. After these tiebacks were installed, it was possible to begin excavation of the temporary sand backfill and the accumulated debris. Tiebacks were similarly installed at the other exterior substructure walls to provide lateral support as the damaged slabs and debris were excavated and removed from the site.

2.3 Observations and Findings

The structural damage sustained by each of the two buildings as a result of the terrorist attacks was massive. The fact that the structures were able to sustain this level of damage and remain standing for an extended period of time is remarkable and is the reason that most building occupants were able to evacuate safely. Events of this type, resulting in such substantial damage, are generally not considered in building design, and the ability of these structures to successfully withstand such damage is noteworthy.

Preliminary analyses of the damaged structures, together with the fact the structures remained standing for an extended period of time, suggest that, absent other severe loading events such as a windstorm or earthquake, the buildings could have remained standing in their damaged states until subjected to some significant additional load. However, the structures were subjected to a second, simultaneous severe loading event in the form of the fires caused by the aircraft impacts.

The large quantity of jet fuel carried by each aircraft ignited upon impact into each building. A significant portion of this fuel was consumed immediately in the ensuing fireballs. The remaining fuel is believed either to have flowed down through the buildings or to have burned off within a few minutes of the aircraft impact. The heat produced by this burning jet fuel does not by itself appear to have been sufficient to initiate the structural collapses. However, as the burning jet fuel spread across several floors of the buildings, it ignited much of the buildings' contents, causing simultaneous fires across several floors of both buildings. The heat output from these fires is estimated to have been comparable to the power produced by a large commercial power generating station. Over a period of many minutes, this heat induced additional stresses into the damaged structural frames while simultaneously softening and weakening these frames. This additional loading and the resulting damage were sufficient to induce the collapse of both structures.

Because the aircraft impacts into the two buildings are not believed to have been sufficient to cause collapse without the ensuing fires, the obvious question is whether the fires alone, without the damage from the aircraft impact, would have been sufficient to cause such a collapse. The capabilities of the fire protection systems make it extremely unlikely that such fires would develop without some unusual triggering event like the aircraft impact. For all other cases, the fire protection for the tower buildings provided in-depth protection. The first line of defense was the automatic sprinkler protection. The sprinkler system was intended to respond quickly and automatically to extinguish or confine a fire. The second line of defense consisted of the manual (FDNY/Port Authority Fire Brigade) firefighting capabilities, which were supported by the building standpipe system, emergency fire department use elevators, smoke control system, and other features. Manual suppression by FDNY was the principal fire protection mechanism that controlled a large fire that occurred in the buildings in 1975. Finally, the last line of defense was the structural fire resistance. The fire resistance capabilities would not be called upon unless both the automatic and manual suppression systems just described failed. In the incident of September 11, not only did the aircraft impacts disable the first two lines of defense, they also are believed to have dislodged fireproofing and imposed major additional stresses on the structural system.

Had some other event disabled both the automatic and manual suppression capabilities and a fire of major proportions occurred while the structural framing system and its fireproofing remained intact, the third line of defense, structural fireproofing, would have become critical. The thickness and quality of the fireproofing materials would have been key factors in the rate and extent of temperature rise in the floor trusses and other structural members. In the preparation of this report, there has not been sufficient analysis to predict the temperature and resulting change in strength of the individual structural members in order to approximate the overall response of the structure. Given the redundancy in the framing system and the capability of that system to redistribute load from a weakened member to other parts of the structural system, it is impossible, without extensive modeling and other analysis, to make a credible prediction of how the buildings would have responded to an extremely severe fire in a situation where there was no prior structural damage. Such simulations were not performed within the scope of this study, but should be performed in the future.

Buildings are designed to withstand loading events that are deemed credible hazards and to protect the public safety in the event such credible hazards are experienced. Buildings are not designed to withstand any event that could ever conceivably occur, and any building can collapse if subjected to a sufficiently extreme loading event. Communities adopt building codes to help building designers and regulators determine those loading events that should be considered as credible hazards in the design process. These building codes are developed by the design and regulatory communities themselves, through a voluntary committee consensus process. Prior to September 11, 2001, it was the consensus of these communities that aircraft impact was not a sufficiently credible hazard to warrant routine consideration in the design of buildings and, therefore, the building codes did not require that such events be considered in building design. Nevertheless, the design of WTC 1 and WTC 2 did include at least some consideration of the probable response of the buildings to an aircraft impact, albeit a somewhat smaller and slower moving aircraft than those actually involved in the September 11 events. Building codes do consider fire as a credible hazard and include extensive requirements to control the spread of fire throughout buildings, to delay the onset of fire-induced structural collapse, and to facilitate the safe egress of building occupants in a fire event. For fire-protected steel-frame buildings, like WTC 1 and WTC 2, these code requirements had been deemed effective and, in fact, prior to September 11, there was no record of the fire-induced-collapse of such structures, despite some very large uncontrolled fires.

The ability of the two towers to withstand aircraft impacts without immediate collapse was a direct function of their design and construction characteristics, as was the vulnerability of the two towers to collapse a result of the combined effects of the impacts and ensuing fires. Many buildings with other design and construction characteristics would have been more vulnerable to collapse in these events than the two towers, and few may have been less vulnerable. It was not the purpose of this study to assess the code-conformance of the building design and construction, or to judge the adequacy of these features. However, during the course of this study, the structural and fire protection features of the buildings were examined. The study did not reveal any specific structural features that would be regarded as substandard, and, in fact, many structural and fire protection features of the design and construction were found to be superior to the minimum code requirements.

Several building design features have been identified as key to the buildings' ability to remain standing as long as they did and to allow the evacuation of most building occupants. These included the following:

- robustness and redundancy of the steel framing system
- adequate egress stairways that were well marked and lighted
- conscientious implementation of emergency exiting training programs for building tenants

Similarly, several design features have been identified that may have played a role in allowing the buildings to collapse in the manner that they did and in the inability of victims at and above the impact floors to safely exit. These features should not be regarded either as design deficiencies or as features that should be prohibited in future building codes. Rather, these are features that should be subjected to more detailed evaluation, in order to understand their contribution to the performance of these buildings and how they may perform in other buildings. These include the following:

- the type of steel floor truss system present in these buildings and their structural robustness and redundancy when compared to other structural systems
- use of impact-resistant enclosures around egress paths
- resistance of passive fire protection to blasts and impacts in buildings designed to provide resistance to such hazards
- grouping emergency egress stairways in the central building core, as opposed to dispersing them throughout the structure

During the course of this study, the question of whether building codes should be changed in some way to make future buildings more resistant to such attacks was frequently explored. Depending on the size of the aircraft, it may not be technically feasible to develop design provisions that would enable all structures to be designed and constructed to resist the effects of impacts by rapidly moving aircraft, and the ensuing fires, without collapse. In addition, the cost of constructing such structures might be so large as to make this type of design intent practically infeasible.

Although the attacks on the World Trade Center are a reason to question design philosophies, the BPS Team believes there are insufficient data to determine whether there is a reasonable threat of attacks on specific buildings to recommend inclusion of such requirements in building codes. Some believe the likelihood of such attacks on any specific building is deemed sufficiently low to not be considered at all. However, individual building developers may wish to consider design provisions for improving redundancy and robustness for such unforeseen events, particularly for structures that, by nature of their design or occupancy, may be especially susceptible to such incidents. Although some conceptual changes to the building codes that could make buildings more resistant to fire or impact damage or more conducive to occupant egress were identified in the course of this study, the BPS Team felt that extensive technical, policy, and economic study of these concepts should be performed before any specific code change recommendations are developed. This report specifically recommends such additional studies. Future building code revisions may be considered after the technical details of the collapses and other building responses to damage are better understood.

2.4 Recommendations

The scope of this study was not intended to include in-depth analysis of many issues that should be explored before final conclusions are reached. Additional studies of the performance of WTC 1 and WTC 2 during the events of September 11, 2001, and of related building performance issues should be conducted. These include the following:

- During the course of this study, it was not possible to determine the condition of the interior structure of the two towers, after aircraft impact and before collapse. Detailed modeling of the aircraft impacts into the buildings should be conducted in order to provide understanding of the probable damage state immediately following the impacts.

- Preliminary studies of the growth and heat flux produced by the fires were conducted. Although these studies provided useful insight into the buildings' behavior, they were not of sufficient detail to permit an understanding of the probable distribution of temperatures in the buildings at various stages of the event and the resulting stress state of the structures as the fires progressed. Detailed modeling of the fires should be conducted and combined with structural modeling to develop specific failure modes likely to have occurred.

- The floor framing system for the two towers was complex and substantially more redundant than typical bar joist floor systems. Detailed modeling of these floor systems and their connections should be conducted to understand the effects of localized overloads and failures to determine ultimate failure modes. Other types of common building framing should also be examined for these effects.

- The fire-performance of steel trusses with spray-applied fire protection, and with end restraint conditions similar to those present in the two towers, is not well understood, but is likely critical to the building collapse. Studies of the fire-performance of this structural system should be conducted.

- Observation of the debris generated by the collapse of the towers and of damaged adjacent structures suggests that spray-applied fireproofing may be vulnerable to mechanical damage from blasts and impacts. This vulnerability is not well understood. Tests of these materials should be conducted to understand how well they withstand such mechanical damage and to determine whether it is appropriate and feasible to improve their resistance to such damage.

- In the past, tall buildings have occasionally been damaged, typically by earthquakes, and experienced collapse within the damaged zones. Those structures were able to arrest collapse before they progressed to a state of total collapse. The two WTC towers were able to arrest collapse from the impact damage, but not from the resulting fires when combined with the impact effects of the aircraft attacks. Studies should be conducted to determine, given the great size and weight of the two towers, whether there are feasible design and construction features that would permit such buildings to arrest or limit a collapse, once it began.

2.5 References

BBC News. 2001. "We Ran for Our Lives." Account of Mike Shillaker. September 13.

Cauchon, D. 2001a. "For Many on Sept. 11, Survival Was No Accident," USAToday.com. December 19.

Cauchon, D. 2001b. "Four Survived by Ignoring Words of Advice," USAToday.com. December 19.

Computers and Structures, Inc. (CSI). 2000. SAP-2000. Berkeley, CA.

Dateline NBC. 2001. "The Miracle of Ladder Company 6." September 28.

Hearst, D. 2001. "Attack on America: Survivors: Suddenly they started to yell out, 'get out now': Bravery and fear mingled with disbelief," *Guardian Home Pages*, page 15. Account of Simon Oliver. September 13.

Labriola, J. 2001. Personal account. Channel 4 News, "Inside the World Trade Center," broadcast. September 13.

Masetti, A. 2001. Personal account received by email. December 21.

Mayblum, A. 2001. Personal account. www.worldtradecenternews.org/survivorstory.html, World Trade Center Miracles section. September 18.

New York Board of Fire Underwriters. 1975. *One World Trade Center Fire, February 13, 1975*.

Nicholson, W. J.; Rohl, A. N.; Wesiman, I.; and Seltkoff, I.J. 1980. *Environmental Asbestos Concentrations in the United States,* page 823. Environmental Sciences Laboratory, Mount Zion Hospital, New York, NY.

Scripps, H. 2001. "I walked out ... I made it out alive," BostonHerald.com. Account of John Walsh. September 14.

Shark, G., and McIntyre, S. December 5, 2001. ABS. Personal account.

Smith, D. 2002. *Report from Ground Zero*. Viking Penguin, New York. p. 29.

Zalosh, R. G. 1995. "Explosion Protection," *SFPE Handbook of Fire Protection Engineering*, 2nd edition. Quincy, MA.

William Baker

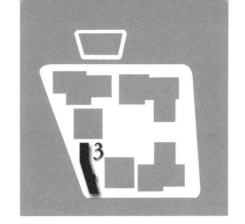

3 WTC 3

3.1 Design and Construction Features

WTC 3 (also known as the World Trade Center Hotel, the Vista Hotel, and the Marriott Hotel) was almost completely destroyed as a result of the September 11 attacks due to debris from the collapse of the adjacent WTC 1 and WTC 2. This chapter describes the structural design and construction features of the building as well as details of its collapse. The information contained in this section is based on architectural and structural design drawings dating to 1978, photographs, written reports and articles, videotape recordings, and interviews with Marriott staff. Steel shop drawings were not available for review.

3.1.1 Project Overview

WTC 3 was a steel-framed hotel building located on the southwest corner of the WTC Complex immediately to the south of WTC 1 (the north tower) and to the west of WTC 2 (the south tower). The site was bounded on the west by West Street and on the south by Liberty Street. The building was designed in 1978/1979 by Skidmore, Owings & Merrill (SOM), architects; Weiskopf & Pickworth, structural engineers; and Jaros, Baum & Bolles, mechanical engineers. Weiskopf & Pickworth designed the superstructure for the hotel. Skilling, Helle, Christiansen, and Robertson designed the transfer system between the hotel column grid and the below grade parking grid as well as the structure below the transfer system. WTC 3 opened in 1981 and housed the 825-room Vista hotel. The original owner and operator of the hotel was Hilton International, but the property was sold in 1996 to Marriott Hotels (Harris 2001). Marriott operated the hotel from 1996 until the attacks on September 11, 2001.

3.1.2 Building Description

The building had 22 stories above grade and 6 stories below grade (labeled B1 through B6.) The roof parapet line was approximately 242 feet above West Street. The north and west elevations are shown in Figure 3-1 and the south and east elevations are shown in Figure 3-2, as taken from the original architectural drawings from SOM. The first level below grade provided loading docks, building services, and other functions for the hotel. The other levels below grade were part of the overall WTC complex and provided parking space. The ground level housed the hotel lobby and ballroom. The two stories above the ground level lobby were occupied by restaurants and conference facilities. The next 18 stories accommodated the approximately 825 guest suites. A health club and mechanical services were located on the top floor of the building.

The building was a long rectangle in plan with an obtuse angle change at approximately one-third of the length of the building from the north. The typical floor of the building was approximately 64 feet wide and 330 feet long, with floor story heights of 9 feet 6 inches. Elevators were located at the east side of the building, roughly centered north to south, directly opposite the building entry, which was located on the west side of the building. The general arrangement of the hotel suites is indicated in Figure 3-3.

CHAPTER 3: WTC 3

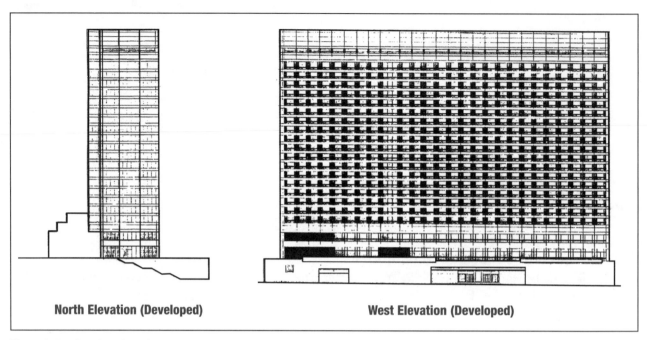

Figure 3-1   Developed north and west elevations (SOM 1979).

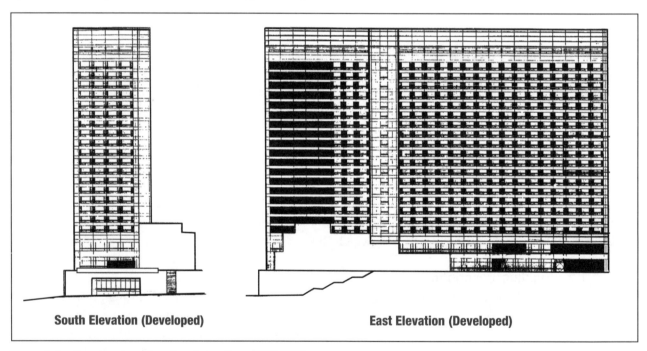

Figure 3-2   Developed south and east elevations (SOM 1979).

3.1.3   Structural Description

The primary structural frame of WTC 3 was composed of rolled, wide-flange structural steel columns, floor beams, and girders. The column grid for the building consisted of approximately twelve 26-foot-wide bays in the north-south direction with non-typical bays at the south end of the building and at the location of the plan angle change. In the east-west direction, there were three bays with column spacings of 18 feet 9-7/8 inches, 22 feet 6 inches, and 18 feet 9-3/4 inches. Steel columns were standard wide-flange W14-series shapes throughout (up to W14x500 at the 2nd floor). Details of column splices were not indicated on the structural design drawings.

The hotel structural system was transferred from the hotel column grid to the substructure grid at different levels. Most of the column transfers occurred right above the lobby level to accommodate the large open space required for the hotel lobby and ballroom. Some column transfers also occurred below the lowest guestroom floor. Skilling, Helle, Christiansen, and Robertson originally designed all the transfers. Following the 1993 bombing, extensive renovations to the hotel required additional transfer girders above the lobby level. These new transfers were designed by Leslie E. Robertson Associates (*Engineering News Record* 1994). Details of the 2nd floor transfer system are not included in this report.

Structural steel was ASTM A36 throughout except for the 2nd floor transfer girders, which were ASTM A572-Grade 50. Typical floor beams were W16 and W21 rolled shapes, spaced 13 feet 0 inches on center. The connection design reactions for the floor beams were specified to be based on the uniformly loaded beam end shear. Typical simple shear connection types are not indicated on the design drawings. Although the structural drawings prepared by Weiskopf & Pickworth indicate typical details of stud shear connectors, the total extent of composite design is unclear. The drawings show that composite action was used all through the 4th floor and on a few elements of the 21st floor. There is no indication that the typical guestroom floors (i.e., 5th through 20th floors) made use of composite action. The design drawings specified that all steel was to be fireproofed, and that all steel pieces weighing less than 28 pounds per linear foot were to have double spray-on thickness for fireproofing. The typical hotel floor framing plan is shown in Figure 3-4, as taken from the original structural design drawings prepared by Weiskopf & Pickworth.

Uniform floor live load allowance was specified to be 40 pounds per square foot (psf) for the typical guest room areas.

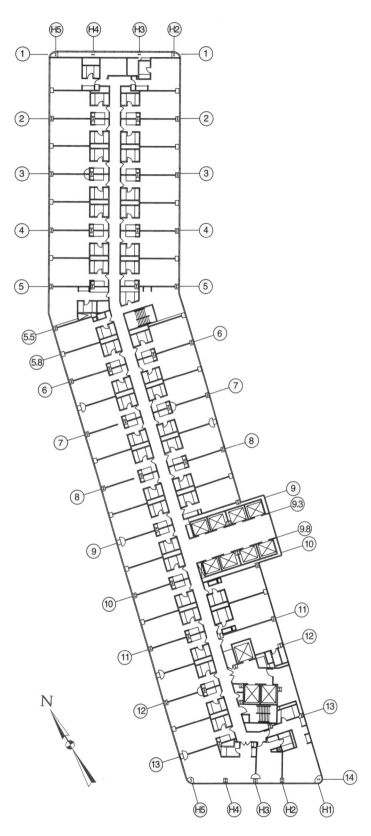

Figure 3-3   Typical hotel floor plan (SOM 1981).

CHAPTER 3: *WTC 3*

Uniform floor dead load allowance included the weight of the floor structure, 10 psf for partitions, and 4 psf for ceilings and finishes. The roof design live load was 30 psf throughout.

The typical guestroom floor slabs spanned 13 feet 0 inches between floor beams and consisted of 3-inch-deep composite steel floor deck (noted as 20 gauge on the original Weiskopf & Pickworth design drawings), with 3-1/4-inch lightweight concrete topping above the top of the steel deck and welded wire fabric reinforcement. The top of the slab was at the same elevation for the guestrooms, restrooms, and corridors. However, in the restrooms and corridors, the floor beams were raised 1 inch to provide increased headroom for service ducts. To accommodate this situation, the steel deck was only 2 inches deep in these areas. The edge of slab was 1 foot 9 inches beyond the column line and was supported by a secondary W14 spandrel beam placed outboard of the perimeter column line.

Resistance to lateral loads in the long (north-south) direction was provided through parallel moment resisting beam/column frames located along all four major column lines. Generally, W16-series girders were utilized for the two interior moment frame lines and W21-series girders for the two exterior moment frame lines. In the short transverse (east-west) direction, resistance to lateral loads was provided through a combination of concentrically and eccentrically diagonally braced frames. These braced frames were located along partition lines between rooms and around the central north-south-running corridor. Each major column line, with the exception of the north and south facades, was diagonally braced (26 feet 0 inches on center). In combination, each column in the building participated in the lateral load resisting system, with most columns involved in resistance along both orthogonal directions. A typical transverse braced frame elevation is shown in Figure 3-5.

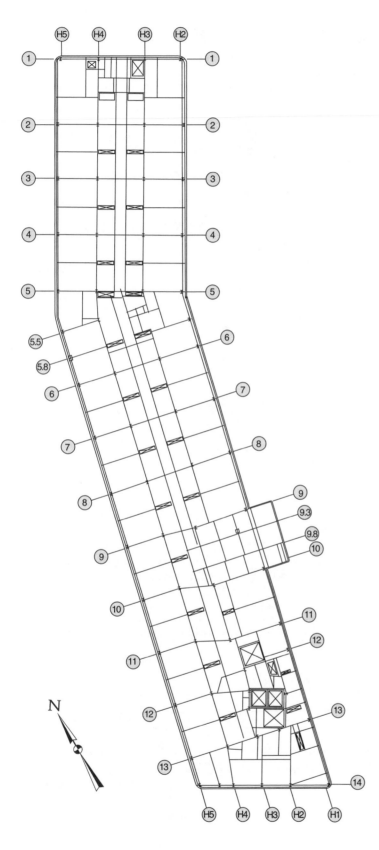

Figure 3-4   Typical hotel floor framing plan (Weiskopf & Pickworth 1979a).

3.2 1993 Attack

WTC 3 was damaged in the February 26, 1993, bombing of the WTC. The bomb was set off in the second basement level parking garage under the north end of the hotel, adjacent to WTC 1. The explosion caused a major collapse of the slab at level B2 (approximately 130 feet by 130 feet in dimension) and a major, but smaller, slab collapse at Level B1 (approximately 50 feet by 80 feet). The West Street level (Concourse level) had a limited collapse (approximately 18 feet by 22 feet). Level 1 (Plaza level) was not ruptured, but had an area (10 feet by 10 feet) that was deflected upward. (The above data were taken from Isner and Klem [1994]). There were no slabs at levels B3 and B4 directly below the blast, and the debris landed on level B5. There was also significant damage to non-loadbearing partition walls and mechanical equipment. The damage was subsequently repaired. The bombing has been more extensively described in several reports, including those by Isner and Klem (1994) and the U.S. Fire Administration (1993).

3.3 2001 Attacks

3.3.1 Fire and Evacuation

The following account was developed through interviews with Marriott Hotel staff.

Small fires on the top floor were ignited as a result of projectiles through the roof, most likely after the impact of the aircraft with WTC 1. At least one of these fires was located in the health club on the top floor. Some jet fuel was reportedly involved in these fires.

Evacuation of the hotel guests and staff was initiated shortly after ignition of the fires. Building occupants were initially directed to the hotel lobby. Later, the building occupants were instructed to evacuate the building. It is unknown whether the fire alarm system was activated in the building. Hotel staff and fire service personnel alerted other building occupants while moving in the corridors on the guest room floors.

All of the building occupants were evacuated from the building. However, two members of the hotel management team had each re-entered the

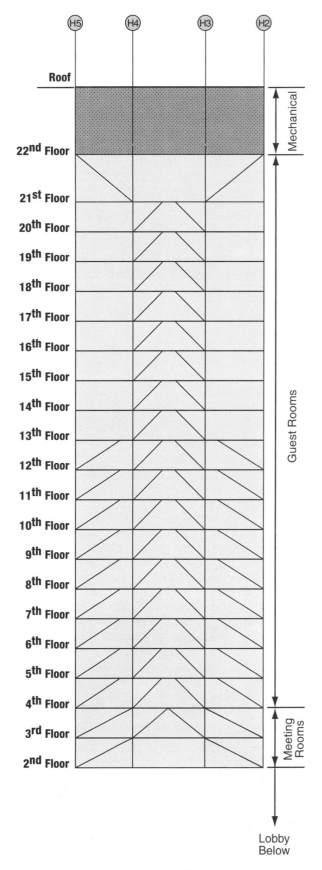

Figure 3-5 Typical transverse bracing elevation (Weiskopf & Pickworth 1979b).

3.3.2 Building Response

The response of WTC 3 to the September 11 events is complex and noteworthy. WTC 3 was subjected to two loading events. The first event involved the collapse of WTC 2, which stood immediately east of WTC 3. Due to its proximity to WTC 2, substantial amounts of debris fell directly on the roof of WTC 3. Figure 3-6 shows large portions of the prefabricated assemblies from WTC 2 falling on top of WTC 3.

Debris from WTC 2 struck the building with sufficient force to crush approximately 16 stories in the center of the building, as shown in Figure 3-7. In spite of this extensive damage, the collapse did not

Figure 3-6
Exterior columns from the collapse of WTC 2 falling on the southern part of WTC 3.

Figure 3-7   Partial collapse of WTC 3 after collapse of WTC 2.

continue down to the foundations or extend horizontally to the edges of the structure. In fact, the two northernmost bays (approximately 60 feet) remained intact all the way to the roof. A similar, but lesser condition existed in the southern bays. Even in the center of the building, the collapse stopped at approximately the 7th floor. This arrested collapse implies that the structure was sufficiently strong and robust to absorb the energy of the falling debris and collapsed floors, but at the same time the connections between the destroyed and remaining framing were able to break apart without pulling down the rest of the structure. This complex behavior resulted in the survival of large portions of the building following the collapse of WTC 2.

The second loading event was the collapse of WTC 1. Debris from WTC 1 fell along the entire length of the hotel. Lower floors at the southwest end of WTC 3 survived although they suffered extensive damage. The remaining portions of the building after both collapses of WTC 1 and WTC 2 are shown in Figure 3-8.

An FDNY fire company was in the building during the collapses of both WTC 1 and WTC 2 and survived. The firefighters were near the top of the building in the process of making sure that there were no civilians present in the building, when the south tower collapsed. Firefighter Heinz Kothe is quoted as saying, "We had no idea what had happened. It just rocked the building. It blew the door to the stairwell open, and it blew the guys up near the door halfway down a flight of stairs. I got knocked down to the landing. The building shook like buildings just don't shake." Subsequently, the firefighters were in the lower portion of the southwest corner of the building when the north tower collapsed (Court 2001).

The Chief Engineer of the Port Authority of New York and New Jersey, Frank Lombardi, was in the lobby of WTC 3 with other Port Authority executives during the collapse of WTC 2. They survived the collapse and were eventually able to leave the building (Rubin and Tuchman 2001).

Figure 3-8   Remains of WTC 3 after collapse of WTC 1 and WTC 2.

3.4 Observations

WTC 3 was subjected to extraordinary loading from the impact and weight of debris from the two adjacent 110-story towers. It is noteworthy that the building resisted both horizontal and vertical progressive collapse when subjected to debris from WTC 2. The overloaded portions were able to break away from the rest of the structure without pulling it down, and the remaining structural system was able to remain stable and support the debris load. The structure was even capable of protecting occupants on lower floors after the collapse of WTC 1.

3.5 Recommendations

WTC 3 should be studied further to understand how it resisted progressive collapse.

3.6 References

Court, Ben, et al. 2001. "The Fire Fighters," *Mens' Journal*. Vol. 10, No. 10, pp. 70. November.

Harris, Bill. 2001. *The World Trade Center, a Tribute*. Courage Books, Philadelphia. November.

Isner, Michael, and Klem, Thomas. 1994. *The World Trade Center Explosion and Fire, February 26, 1993*. Fire Investigation Report. National Fire Protection Association.

Post, Nadine. 1994. "Much Done, More to Come." *Engineering News Record*. Vol. 232, No. 9. February 28.

Rubin, D., and Tuchman, J. 2001. "WTC Engineers Credit Design in Saving Thousands of Lives." *Engineering News Record.* Vol. 247, No. 16, pp. 12. October 15.

Skidmore, Owings and Merrill, LLP. 1979a. Drawing A-11, WTC Hotel. New York, NY.

Skidmore, Owings and Merrill, LLP. 1979b. Drawing A-12, WTC Hotel. New York, NY.

Skidmore, Owings and Merrill, LLP. 1981. Drawing A-6C. New York, NY.

United States Fire Administration. 1993. *The World Trade Center Bombing: Report and Analysis.* Technical Reports Series. Prepared in association with the Federal Emergency Management Agency. October.

Weiskopf &Pickworth. 1979a. Drawing S-5. New York, NY.

Weiskopf &Pickworth. 1979b. Drawing S-11. New York, NY.

Jonathan Barnett
Richard Gewain
Ramon Gilsanz
Harold "Bud" Nelson

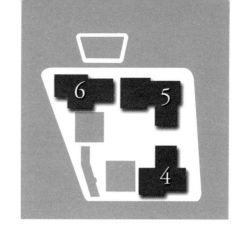

4 WTC 4, 5, and 6

4.1 Design and Construction Features

WTC 4, 5, and 6 are eight- and nine-story steel-framed office buildings, located on the north and east sides of the WTC Plaza, that were built circa 1970. The buildings had a range of occupancies, including standard office and retail space. There were underground parking facilities and access to the WTC Concourse, as well as the Port Authority Trans-Hudson (PATH) and New York City subway system.

Because of their close proximity to WTC 1 and WTC 2, all three buildings were subjected to severe debris impact damage when the towers collapsed, as well as the fires that developed from the debris. Most of WTC 4 collapsed when impacted by the exterior column debris from WTC 2; the remaining section had a complete burnout. WTC 5 and WTC 6 were impacted by exterior column debris from WTC 1 that caused large sections of localized collapse and subsequent fires spread throughout most of the buildings. All three buildings also were able to resist progressive collapse, in spite of the extensive local collapses that occurred.

This chapter describes the design and construction features of these buildings and observed damages. Site observations of damage in WTC 5 and WTC 6 were conducted by team members, although access in WTC 6 was severely limited. WTC 4 was declared unsafe, and no access was allowed.

All three buildings were designed by Minoru Yamasaki & Associates, Architects; Emery Roth & Sons, Architects; and Skilling Helle Christiansen Robertson, Structural Engineers. The buildings had similar design features, but their configurations were somewhat different. Therefore, because most site observations were made in WTC 5, the following discussion focuses primarily on this building, and is assumed to be applicable to all three structures.

4.1.1 Structural Design Features

WTC 5 was located in the northeast corner of the WTC Plaza. The nine-story building was L-shaped in plan, with overall dimensions of 330 feet by 420 feet, providing approximately 120,000 square feet per floor (Figure 4-1). Floors were constructed of 4-inch-thick lightweight concrete fill on metal deck (with a combined thickness of 5-1/2 inches), supported by structural steel framing. The steel floor frame had shear studs welded to the top flange to create a composite floor system with the concrete deck. Wide-flange structural columns were placed on a regular 30-foot-square grid pattern. The floor plates cantilevered out 15 feet from the exterior column lines on all sides. To support this cantilever and provide the basic lateral resistance for the structure, a pair of W27 wide-flange beams were provided at each column line. These doubled wide-flange beams extended between the two outermost column lines, forming a moment-resisting frame, and cantilevered past the outer columns.

Floor beams were typically W16 wide-flange members. At interior column lines, a column-tree system was used, in which a 4-foot-long stub was shop-welded to the column on each side, and the floor girder was

CHAPTER 4: WTC 4, 5, and 6

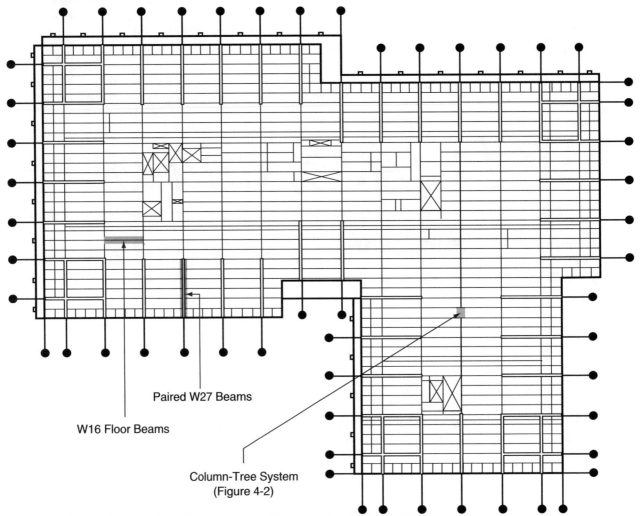

Figure 4-1   Typical floor plan for WTC 5 (Worthington, Skilling, Helle & Jackson 1968).

simply connected, with shear tabs, to the cantilevers (Figure 4-2). Floor 9 and roof level framing was conventional for steel-frame construction and did not include a column-tree system, as illustrated in Figure 4-3.

4.1.2 Fire Protection Features

WTC 5 appeared to have typical combustible contents for an office building, including furnishings, paper, etc. No evidence of any other type of fire load was noted. There appeared to be local concentrations of heavier fire loads, such as file storage, in some areas of the floors that exceeded the average combustible fire load normally associated with office occupancies. A raised sub-floor was present in a portion of the 6th floor, indicative of a computer room or electronic equipment area.

At the time of the September 11 attacks, WTC 5 was equipped with an automatic sprinkler system. The columns had the characteristics of a 3-hour fire resistance rating. The floor assembly (floor and floor beams) had the characteristics of a 2-hour fire resistance rating. Typical fire resistance ratings of roofs of this type are 1-1/2 hours. The exterior non-loadbearing walls did not appear to have a fire resistance rating.

Based on the building plans and on-site observations of WTC 5, the fire protection material for the steel members was a spray-applied mineral fiber applied directly to the steel columns, beams, and girders. Sprayed fiber is a commonly used low-density material.

CHAPTER 4: *WTC 4, 5, and 6*

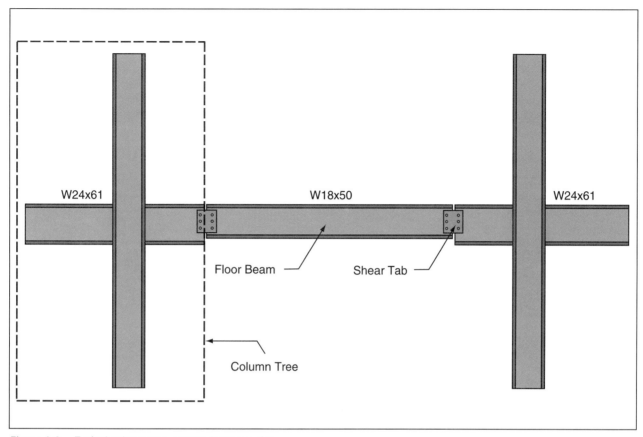

Figure 4-2   Typical column-tree system (not to scale).

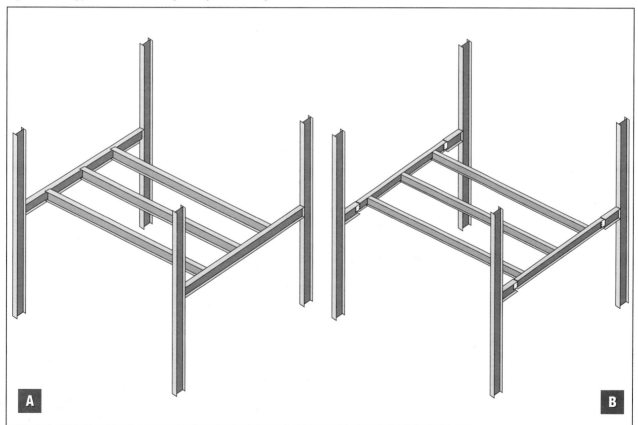

Figure 4-3   Typical interior bay framing in WTC 5. (A) Floor 9 and roof level. (B) Floors 4, 5, 6, 7, and 8.

CHAPTER 4: WTC 4, 5, and 6

There was one continuous open escalator from the Concourse level to the 4th floor and one open escalator connected the Plaza level with the mezzanine. Four stairways connected the Plaza level to the 8th floor, and three of those stairways continued up to the 9th floor.

The stairway enclosure core areas were constructed of two layers of 5/8-inch-thick Type-X (fire resistant) gypsum wallboard on both sides of steel studs. The elevator shafts were constructed of HT shaped steel studs and gypsum wallboard and coreboard, which provides a 2-hour fire resistance rating. This type of wall framing permits construction of elevator shafts from the office side of the wall. These core areas are illustrated in Figure 4-4.

4.2 Building Loads and Performance

There was major impact to WTC 4 from the collapse of one or both of the WTC towers (most likely WTC 2) that destroyed all but the northern 50 feet of the building. Extensive damage is evident in Figure 4-5.

WTC 5 was damaged by impact and subsequent fires. The impact damage areas in WTC 5 are shown in Figure 4-6. The debris damage caused localized collapses from the roof to the 3rd floor in most of the areas where exterior columns impacted the structure. Ensuing fires that burned unchecked in the building caused a localized collapse from the 9th floor to the 4th floor. Figure 4-7 diagrammatically shows the damaged and collapsed areas of WTC 5 due to impact and fire.

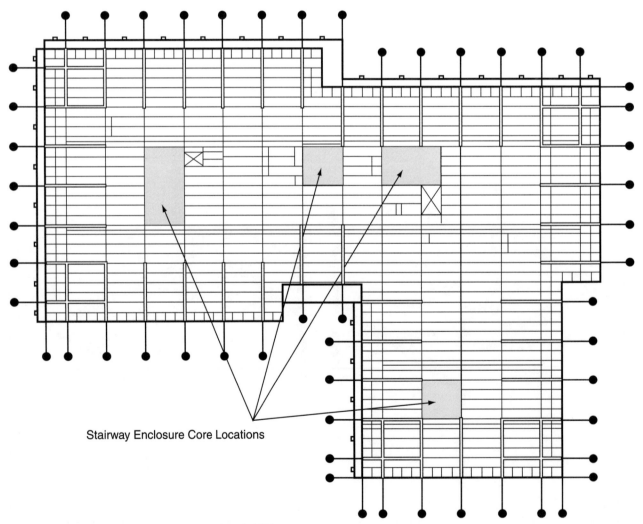

Figure 4-4 Stairway enclosure core locations in WTC 5.

CHAPTER 4: *WTC 4, 5, and 6*

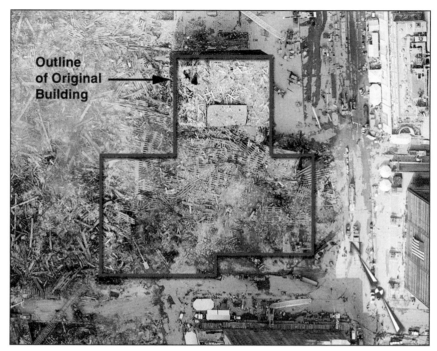

Figure 4-5   Damage to WTC 4.

Figure 4-6   Damage to WTC 5.

FEDERAL EMERGENCY MANAGEMENT AGENCY

CHAPTER 4: *WTC 4, 5, and 6*

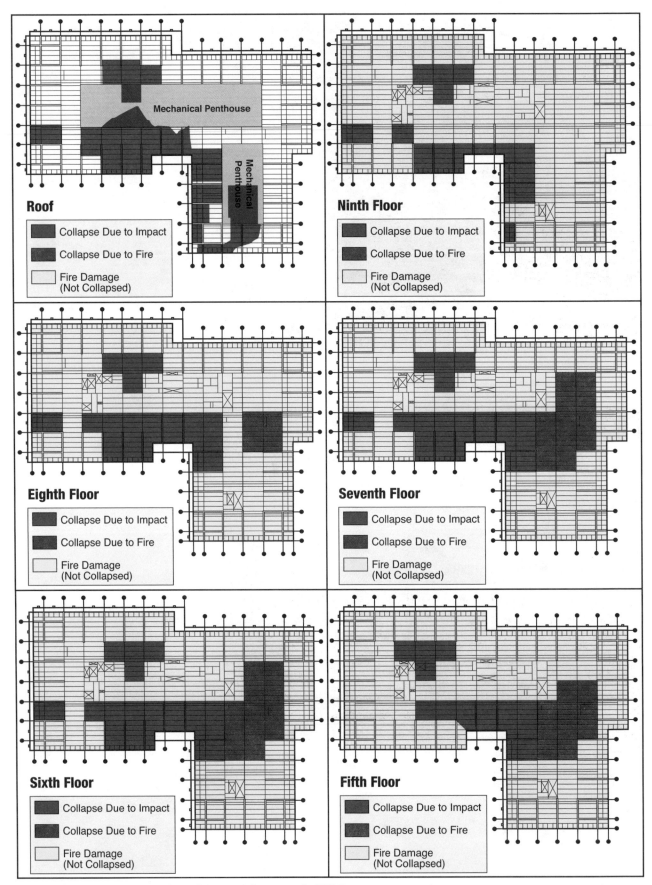

Figure 4-7   Approximate locations of damaged floor areas in WTC 5.

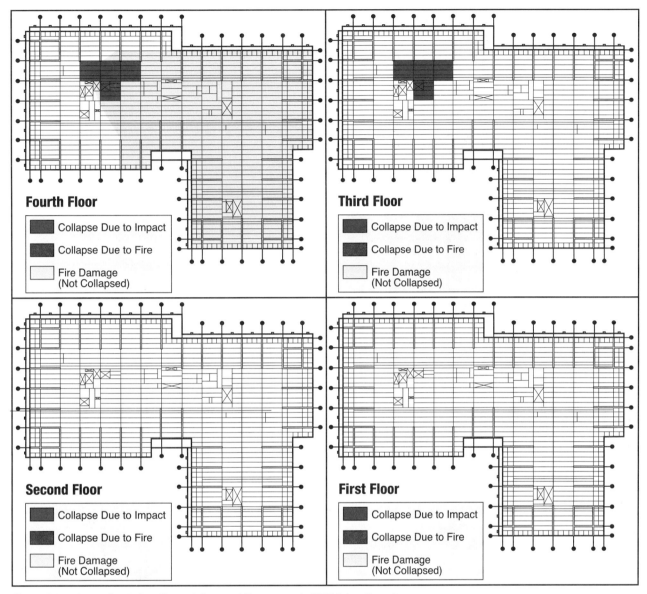

Figure 4-7   Approximate locations of damaged floor areas in WTC 5 (continued).

WTC 6 suffered significant impact and fire damage from the collapse of WTC 1, as shown in Figure 4-8. Most of the impact appears to have been in the center of the building, where the damage extended to the ground level. Figures 4-9 and 4-10 illustrate the magnitude of the damage.

4.2.1   Impact Damage to WTC 5

The impact was most severe at the inside corner at the junction of the north-south and east-west portions of the L-shaped building, as well as at a limited region in the west region of the north-south portion.

The debris impact caused partial collapses of the roofs and some floors beneath the points of impact in all three buildings. In WTC 5, it caused partial collapses down to floor 3. The debris impact also ignited fires that spread throughout the building.

Many areas of buckled steel-beam flanges appeared to have been caused by debris impact. There were 3-1/2 inch steel pipe façade supports along the building perimeter, many of which had buckled from taking part of the collapsed floor loads above the pipe supports (see Figure 4-11). In areas that locally collapsed from

CHAPTER 4: *WTC 4, 5, and 6*

Figure 4-8   Damage to WTC 6. Note the edge of WTC 5 on the right hand side.

Figure 4-9   Impact damage to WTC 6.

4-8   **WORLD TRADE CENTER BUILDING PERFORMANCE STUDY**

impact, some of the floor beams separated from the floor deck and others separated at the welded connection.

Significant impact damage to WTC 5 is shown in Figure 4-12. The damage was concentrated on the west side of the building.

4.2.2 Fire Damage

Figure 4-13 shows WTC 5 on fire. The source of ignition has not been identified. It is likely that it was due to flaming debris entering the building from WTC 1 and WTC 2. There was a complete burnout of all combustibles from the 5th floor and above. Some steel beams supporting the roof were deformed due to the heat, as illustrated in Figure 4-14, and some local buckling occurred as well. Roof tar entered the floor through the drains. There is no indication that this roof tar played a major role in the fires. One area below the roof at the 8th floor collapsed onto the 7th floor and then both onto the 6th, and so on, down to the 4th floor.

The structural damage due to the fires closely resembled that commonly observed in test assemblies exposed to the ASTM E119 Standard Fire Test. After testing, the deformed shapes of beams, girders, and columns are similar to the structural damage that occurred

Figure 4-10 Impact damage to the exterior façade of WTC 6.

in these buildings. The damage also resembled the fire damage associated with the fire incident at the unsprinklered One Meridan Plaza, a steel-framed building in Philadelphia, and damage observed at experiments conducted at Cardington by the Building Research Establishment (BRE) and British Steel in 1995. These fires are discussed in greater detail in Appendix A, Section A.3.1.3.

Discrete sections of the steel framing were warped or twisted. There was no evidence of weakened connections in the areas of the building that were inspected. Some studs were missing and others were still in place in some areas, even in floor sections that had collapsed. In many deformed beams, there was no evidence of damage at openings cut into the beams, suggesting that web penetration reinforcement design worked as intended, although some localized buckling in beams and girders was observed throughout the burned-out regions.

There was significant fire damage on floors 4 through 8. On some floors, the interior had been completely gutted by fire; on others, the fire damage was severe, but there was still evidence of office partition frames and other light-gauge metal products, except for the 6th floor, which suffered near

Figure 4-11 WTC 5 façade damage.

complete destruction. Even the mid-height partitions were destroyed and had collapsed throughout much of the 6th floor. This level of damage was not evident on the other floors.

The sprinkler system appears not to have operated at all. This was evident due to the lack of water damage throughout, but especially in the lower level bookstore shown in Figure 4-15. Many sprinkler heads were damaged; some even melted and fell off the sprinkler piping in the fire areas.

As illustrated in Figure 4-16, the interior of the exit stair tower at the southwest corner was practically untouched by the fire. There were no burn marks or smoke damage in this exit tower, or on the "safe" or stairwell side of the fire door. At one location, a piece of paper was found taped to the exit stairwall just inside the fire door with no evidence of smoke or charring from fire.

4.3 Analysis of Building Performance

4.3.1 Steel and Frame Behavior

The punctures in the building envelope of WTC 5, caused by falling debris from the WTC towers, are a result that would be expected from large debris falling over a great distance. Generally, the debris punched through the roof and floor slabs and severed or otherwise damaged framing until sufficient energy had dissipated to arrest the collapse.

The debris from the towers caused damage to the outside wall steel framing of WTC 5, but this damage did not cause any additional collapse of the floors. In fact, the steel pipe façade supports (mullions) provided structural redundancy to the floor framing and redistributed some portion of the cantilevered floors to other levels.

CHAPTER 4: *WTC 4, 5, and 6*

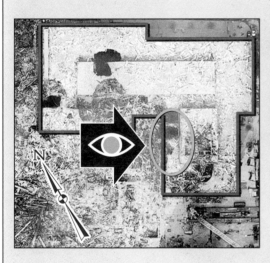

Figure 4-12   Impact damage to WTC 5.

CHAPTER 4: *WTC 4, 5, and 6*

Figure 4-12 Impact damage to WTC 5 (continued).

As illustrated in Figure 4-17, there was local buckling of interior columns. This buckling was most likely due to a combination of fire-induced reductions in strength and a possible increase in stress due to restrained thermal expansion. A detailed explanation of these issues is presented in Appendix A, Section A.3.1.4.

4.3.2 WTC 5 - Local Collapse Mechanisms

Two areas in WTC 5 experienced local collapse under an intact portion of the roof. Although there was debris impact near this area, the symmetrical nature of the collapse strongly suggests that the failures were due to the uncontrolled fires. This is supported by the observation that the columns in this area remained straight and freestanding (see Figure 4-18). This local collapse appeared to have begun at the field connection where beams were connected to shop-fabricated beam stubs and column assemblies as illustrated in Figures 4-19, 4-20, and 4-21.

The structural collapse appeared to be due to a combination of excessive shear loads on bolted connections and unanticipated tensile forces resulting from catenary sagging of the beams. The existence of high shear loads, likely due to collapsing floor loads from above, was evident in many of the column-tree

CHAPTER 4: *WTC 4, 5, and 6*

Figure 4-13   WTC 5 on fire.

Figure 4-14   Deformed beams in WTC 5.

FEDERAL EMERGENCY MANAGEMENT AGENCY

Figure 4-15 Unburned bookstore in WTC 5.

Figure 4-16 Looking through the door into the undamaged stair tower in WTC 5.

Figure 4-17
Buckled beam flange and column on the 8th floor of WTC 5 that was weakened by fire.

beam stub cantilevers that formed diagonal tension field mechanisms in the cantilever webs and plastic moments at the column, as seen in Figure 4-18.

It is apparent that fire weakened the steel, contributing to the large shear-induced deformations observed in several of the cantilever beams. The shear failures observed at connection ends in several of the beam web samples shown in Figure 4-18 are indicative of the tensile forces that developed. The end bearing resistance of the beam web was found to be less than the double shear strength of the high-strength bolts, based on the analysis presented in Appendix B.

Steel framing connection samples were recovered from floors 6, 7, and 8 of WTC 5 with the aid of the New York Department of Design and Construction (DDC) and are described in Figure 4-22. These samples have not been analyzed and are being preserved for future study. The photographs of connection samples in Figure 4-22 indicate that the deformed structure subjected the bolted shear connection to a large tensile force. At 550 °C (1,022 °F), the ultimate resistance of the three bolts is about 45 kips. The capacity increases to approximately 90 kips at room temperature. Connection failure likely occurred between these bounds.

CHAPTER 4: WTC 4, 5, and 6

Figure 4-18 Internal collapsed area in WTC 5.

Tensile catenary action of floor framing members and their connections has been neither a design requirement nor a design consideration for most buildings. Further study of such mechanisms for member failures in fires should be conducted to determine whether current design parameters are adequate for performance under fire loads.

4.4 Observations and Findings

All three buildings suffered extensive fire and impact damage and significant partial collapse. The condition of the stairways in WTC 5 indicates that, for the duration of this fire, the fire doors and the fire protective covering on the walls performed well. There was, however, damage to the fire side of the painted fire doors, and the damage-free condition on the inside or stairwell side of those same doors indicates the doors performed as specified for the fire condition that WTC 5 experienced. These stairway enclosures were unusual for buildings that have experienced fire because they were not impacted by water from firefighting operations. In addition, the stairway doors were not opened during the fire and remained latched and closed throughout the burnout of the floors. Therefore, general conclusions regarding the effectiveness of this type of stairway construction may not be warranted.

The steel generally behaved as expected given the fire conditions in WTC 5. Many beams developed catenary action as illustrated in Figure 4-14. Some columns buckled, as shown in Figure 4-17. The one exception is the limited internal structural collapse in WTC 5. The fire-induced failure that led to this collapse was unexpected. As in the rest of the building, the steel beams were expected to deflect significantly, yet carry the load. This was not the case where the beam connections failed. The failure most likely occurred during the heating of the structure because the columns remained straight and freestanding after the collapse.

Figure 4-19
Internal collapsed area in WTC 5.

The structural redundancy provided by the exterior wall pipe columns helped to support the cantilevered floors. This was important because it kept the cantilevers from buckling near the columns as might be expected.

The limited structural collapse in WTC 5 due to fire impact as described in Section 4.3.2 appeared to be caused by a combination of excessive shear loads and tensile forces acting on the simple shear connections of the infill beams. The existence of high shear loads was evident in many of the column tree beam stub cantilevers that formed diagonal tension field failure mechanisms in the cantilever webs, as seen in Figure 4-19.

The end bearing resistance of the beam web was less than the double shear strength of the high-strength bolts. An increased edge distance might have prevented this collapse by increasing the connections' tensile strength. The failure most likely began on the 8th floor and progressed downward, because the 9th floor did not collapse. The 4th floor and those below remained intact.

The 7th floor framing was shop-coated. In some locations, the paint appeared to be in good condition and not discolored by the fire. Paint usually blisters and chars when heated to temperatures of about 100 °C (212 °F). This indicates that the fire protection material remained on the steel during the early phase of the fire and may have fallen off relatively later in the fire as the beams twisted, deflected, and buckled. Additional measures for proper adhesion may be required when applying spray-on fire protection to painted steel.

CHAPTER 4: WTC 4, 5, and 6

Figure 4-20 Internal collapsed area in WTC 5.

Figure 4-21 Internal collapsed area in WTC 5 with closeup of connection failure at column tree.

CHAPTER 4: *WTC 4, 5, and 6*

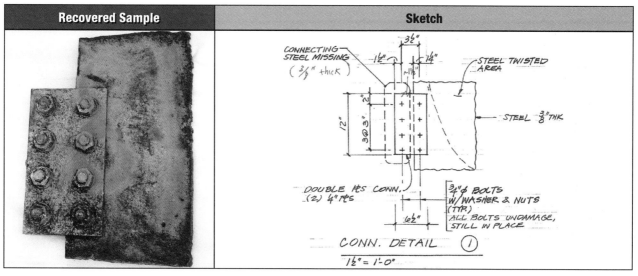

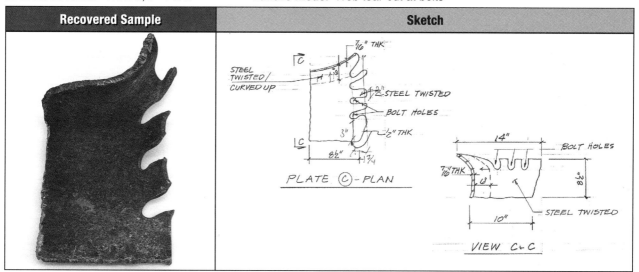

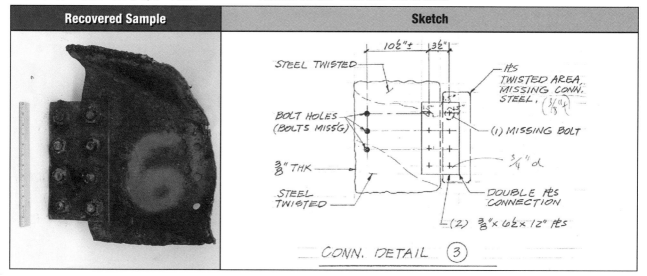

Figure 4-22 Connection samples.

CHAPTER 4: WTC 4, 5, and 6

Recovered From: WTC 5, 6th Floor **Failure Mode:** Web tear-out at bolts

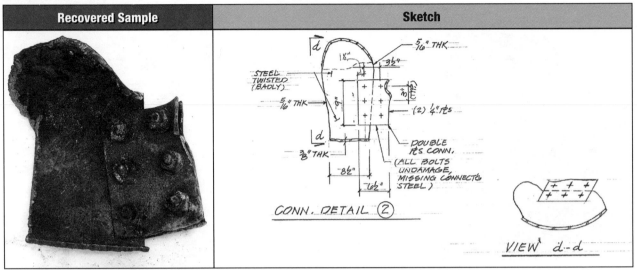

Recovered From: WTC 5, 8th Floor **Failure Mode:** Plate tear-out at bolts

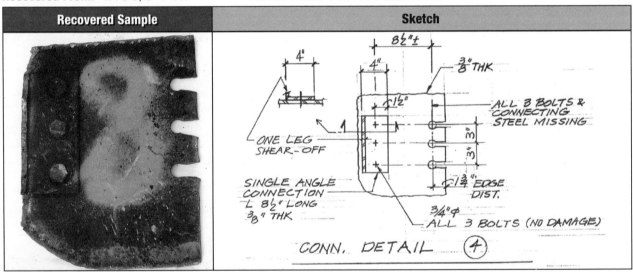

Recovered From: WTC 5, 7th Floor **Failure Mode:** Connection web plate or splice block shear

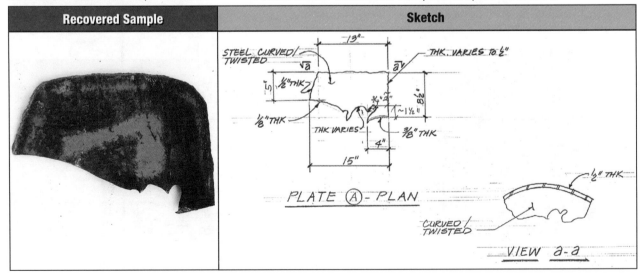

Figure 4-22 Connection samples (continued).

CHAPTER 4: *WTC 4, 5, and 6*

Recovered From: WTC 5, Floor undetermined   **Failure Mode:** Column buckling due to heat exposure

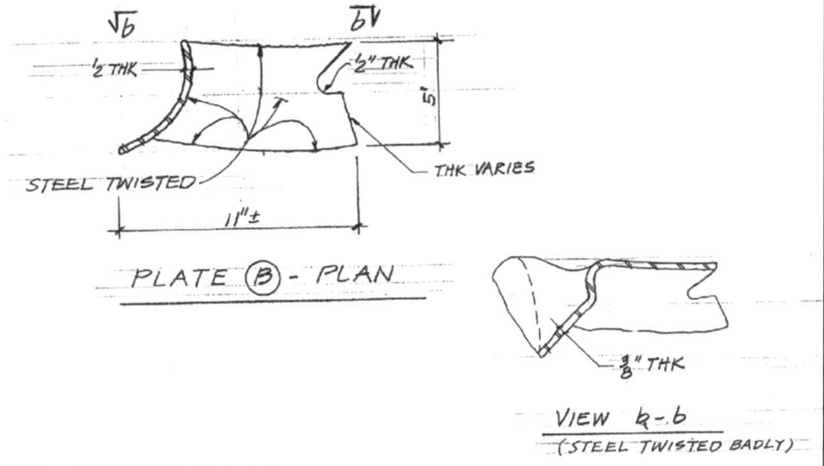

Figure 4-22 Connection samples (continued).

On the lower floors, the steel beams appeared to have heat damage from direct fire impact and there was little or no evidence of shop painting, indicating that fireproofing material was either missing before the fire or delaminated early in the fire exposure.

In general, the buildings responded as expected to the impact loadings. Collapse was often localized, although half of WTC 4 and most of the central part of WTC 6 suffered collapse on all floors. The damage was consistent with the observed impact load.

Reinforced web openings in steel beams performed well, as no damage or local buckling was observed at these locations.

The automatic sprinkler system did not control the fires. Some sprinkler heads fused, but there was no evidence of significant water damage, due to a lack of water. This is consistent with the lack of water damage in the bookstore on the lower level and the complete burnout of the upper floors.

4.5 Recommendations

The scope of this study and the limited time allotted prevented in-depth study of many issues that should be explored before final conclusions are reached. Additional studies of the performance of WTC 4, 5, and 6 during the events of September 11, 2001, and related building performance issues should be conducted. These include the following:

- There is insufficient understanding of the performance of connections and their adequacy under real fire exposures as discussed in Appendix A. This is an area that needs further study. The samples discussed in Section 4.3.2 should be useful in such a study.

- A determination of the combined structural and fire properties of the critical structural connections should be made to permit prediction of their behavior under overload conditions. This can be accomplished with a combination of thermal transfer modeling, structural finite element modeling (FEM), and full-scale physical testing.

4.6 References

AISC. 2001. *Manual of Steel Construction*, LRFD, 3rd Edition. Chicago.

ASTM. 2000. Standard Test Methods for Fire Tests of Building Construction and Materials. ASTM E119. West Conshohocken, PA.

Smith, Dennis. 2002. *Report from Ground Zero: The Story of the Rescue Efforts at the World Trade Center.* Viking Press.

Worthington, Skilling, Helle & Jackson. 1968. "The World Trade Center, Northeast Plaza Building." Structural drawings.

Zalosh, R. G. 1995. "Explosion Protection," *SFPE Handbook of Fire Protection Engineering*, 2nd Edition. Quincy, MA.

Ramon Gilsanz
Edward M. DePaola
Christopher Marrion
Harold "Bud" Nelson

5 WTC 7

5.1 Introduction

WTC 7 collapsed on September 11, 2001, at 5:20 p.m. There were no known casualties due to this collapse. The performance of WTC 7 is of significant interest because it appears the collapse was due primarily to fire, rather than any impact damage from the collapsing towers. Prior to September 11, 2001, there was little, if any, record of fire-induced collapse of large fire-protected steel buildings.

The structural design and construction features of this building, potential fuel loads, fire damage, and the observed sequence of collapse are presented to provide a better understanding of what may have happened. However, confirmation will require additional study and analysis. Information about the structural design and construction features and the observed sequence of events is based upon a review of structural drawings, photographs, videos, eyewitness reports, and a 1986 article about the construction features of WTC 7 (Salvarinas 1986). In addition, the following information and data were obtained from the indicated sources:

- Annotated floor plans and riser diagrams of the emergency generators and related diesel oil tanks and distribution systems (Silverstein Properties 2002)
- Engineering explanation of the emergency generators and related diesel oil tanks and distribution systems (Flack and Kurtz, Inc. 2002)
- Information on the continuity of power to WTC 7 (Davidowitz 2002)
- Summary of diesel oil recovery and spillage (Rommel 2002)
- Information on WTC 7 fireproofing (Lombardi 2002)
- Information on the New York City Office of Emergency Management (OEM) tanks at WTC 7 (Odermatt 2002)

The 47-story office building had 1,868,000 square feet of office space. The top 40 stories of the building (floors 8 to 47) were office type occupancies. Table 5.1 lists the larger tenants of WTC 7. WTC 7 was completed in 1987 by a development team composed of the following parties:

- Owner/Developer: Seven World Trade Company, Silverstein Development Corporation, General Partner
- Construction Manager: Tishman Construction Corporation of New York
- Design Architect: Emery Roth & Sons, P.C.
- Structural Consultant: The Office of Irwin G. Cantor, P.C.
- Mechanical/Electrical Consultant: Syska & Hennessy, P.C.
- Structural Consultant (Con Ed Substation): Leslie E. Robertson Associates

CHAPTER 5: WTC 7

As shown in Figure 1-1 (WTC site map in Chapter 1), WTC 7 was located north of the main WTC complex, across Vesey Street, and was linked to the WTC Plaza by two pedestrian bridges: the large Plaza bridge and a smaller, glass-enclosed pedestrian bridge. The bridges spanned 95 feet across Vesey Street, connecting the Plaza and the 3rd floor of WTC 7. In addition to the office occupancies, WTC 7 also contained an electrical substation, and the WTC Complex shipping ramp, as shown in Figure 5-1.

The substation and shipping ramp occupied major portions of the WTC 7 site. The substation was built prior to the office tower, supplied electrical power to lower Manhattan, and covered approximately half the site. The shipping ramp (5,200 square feet in area, approximately 10 percent of the WTC 7 site) was used by the entire WTC complex.

Table 5.1 WTC 7 Tenants

Floor	Tenant
46-47	Mechanical Floors
28-45	Salomon Smith Barney (SSB)
26-27	Standard Chartered Bank
25	Internal Revenue Service (IRS)
	Department of Defense (DOD)
	Central Intelligence Agency (CIA)
24	Internal Revenue Service (IRS)
23	Office of Emergency Management (OEM)
22	Federal Home Loan Bank of New York
21	First State Management Group
19-21	ITT Hartford Insurance Group
19	National Association of Insurance Commissioners (NAIC) Securities Valuation Office
18	Equal Employment Opportunity Commission (EEOC)
14-17	Vacant
13	Provident Financial Management
11-13	Securities and Exchange Commission
9-10	U.S. Secret Service
7-8	American Express Bank International
7 part	OEM generators and day tank
6	Switchgear, storage
5	Switchgear, generators, transformers
4	Upper level of 3rd floor lobby, switchgear
3	Lobby, SSB Conference Center, rentable space, management offices
2	Open to 1st floor lobby, transformer vault upper level, upper level switchgear
1	Lobby, loading docks, existing Con Ed transformer vaults, fuel storage, lower level switchgear

CHAPTER 5: *WTC 7*

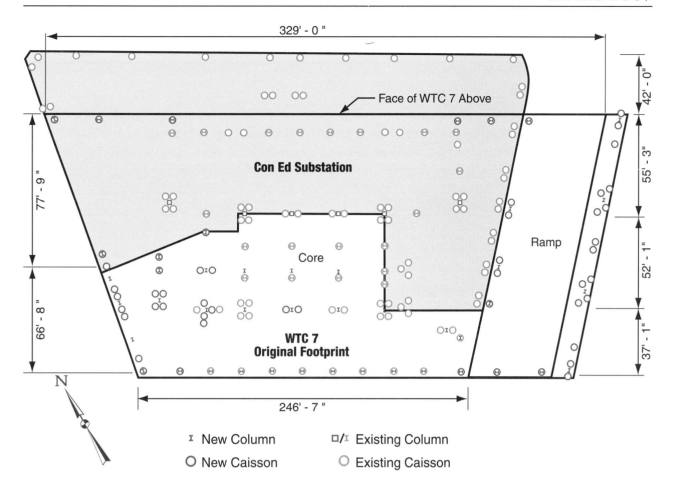

Figure 5-1 Foundation plan – WTC 7.

5.2 Structural Description

5.2.1 Foundations

With the development of an office tower in mind, the Port Authority of New York and New Jersey (hereafter referred to as the Port Authority) installed caissons intended for future construction. However, Seven World Trade Company, Silverstein Development Corporation, General Partner, decided to construct a building much larger in both height and floor area. The designers combined the existing caissons inside the substation with new caissons inside and outside the substation to create the foundation for WTC 7. Figure 5-1 shows the location of pre-existing caissons built when the Con Ed substation was constructed along with new caissons that were installed for the support of the building. The discrepancy in the column locations between the substation and the office tower required transfers to carry loads from the office tower to the substation and finally into the foundation. Old and new caissons, as well as old and new columns, also can be seen in the foundation plan shown in Figure 5-1.

5.2.2 Structural Framing

The typical floor framing shown in Figure 5-2 was used for the 8th through the 45th floors. The gravity framing consisted of composite beams (typically W16x26 and W24x55) that spanned from the core to the perimeter. The floor slab was an electrified composite 3-inch metal deck with 2-1/2-inch normal-weight concrete fill spanning between the steel beams. Below the 8th floor, floors generally consisted of formed slabs with some limited areas of concrete-filled metal decks. There were numerous gravity column transfers, the more significant of these being three interior gravity column transfers between floors 5 to 7 and eight cantilever column transfers in the north elevation at the 7th floor. The column transfers in the exterior walls are shown in the bracing elevations (Figure 5-3).

The lateral load resisting system consisted of four perimeter moment frames, one at each exterior wall, augmented by two-story belt trusses between the 5th and 7th floors and between the 22nd and 24th floors. There were additional trusses at the east and west elevations below the 7th floor. An interior braced core extended from the foundation to the 7th floor. The horizontal shear was transferred into the core at the 5th and the 7th floors. The 5th floor diaphragm (plan shown in Figure 5-4) consisted of a reinforced concrete 14-inch-thick slab with embedded steel T-sections. The 7th floor was an 8-inch-thick reinforced concrete slab.

The 5th and 7th floors contained the diaphragm floors, belt trusses, and transfer girders. A 3-D rendering of Truss 1, Truss 2, Truss 3, and several cantilever transfer girders is shown in Figure 5-5.

5.2.3 Transfer Trusses and Girders

The transfer trusses and girders, shown in Figure 5-6, were located between the 5th and 7th floors. The function and design of each transfer system are described below.

Truss 1 was situated in the northeast sector of the core, and spanned in the east-west direction. As shown in Figure 5-7, this truss was a two-story double transfer structure that provided load transfers between nonconcentric columns above the 7th floor to an existing column and girder at the 5th floor. The girder then provided a second load transfer to an additional two columns. The 7th floor column supported 41 floors and part of the east mechanical penthouse. Its load was transferred through the triangular truss into a column located above an existing substation column and girder at the 5th floor. The 36.5-ton built-up double web girder spanned in the north-south direction between two new columns that started at the foundation and terminated at the 7th floor. The truss diagonals were W14 shapes and the horizontal tie was a 22-ton, built-up shape.

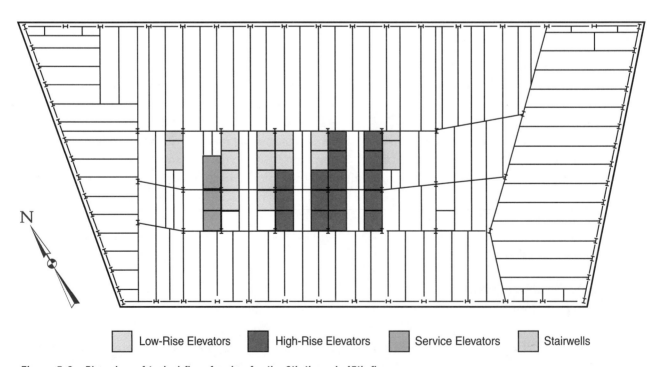

Figure 5-2 Plan view of typical floor framing for the 8th through 45th floors.

CHAPTER 5: *WTC 7*

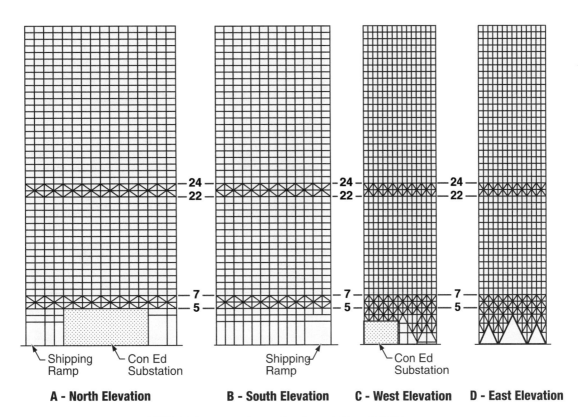

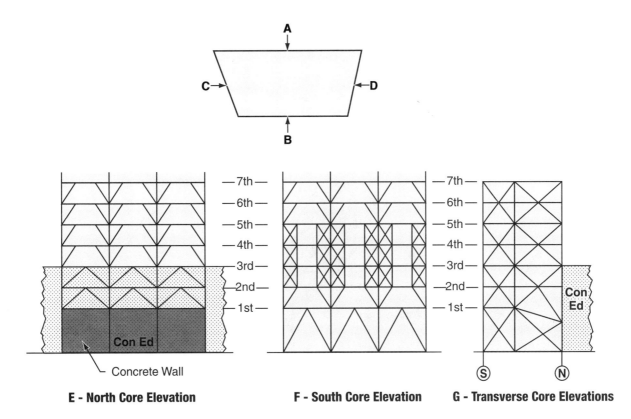

Figure 5-3   Elevations of building and core area.

CHAPTER 5: WTC 7

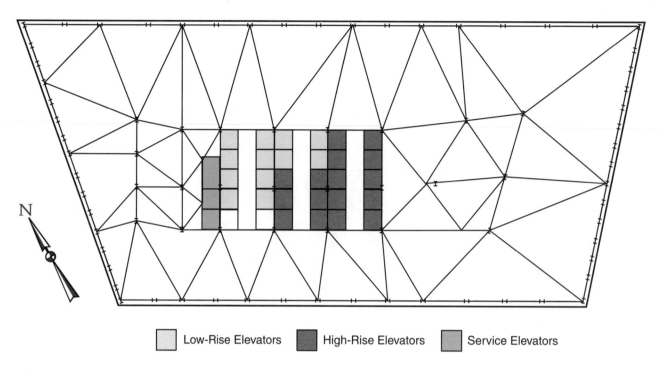

Figure 5-4   Fifth floor diaphragm plan showing T-sections embedded in 14-inch slab.

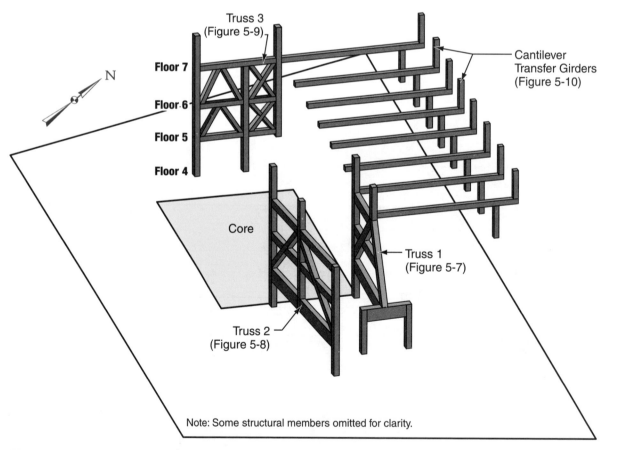

Figure 5-5   3-D diagram showing relation of trusses and transfer girders.

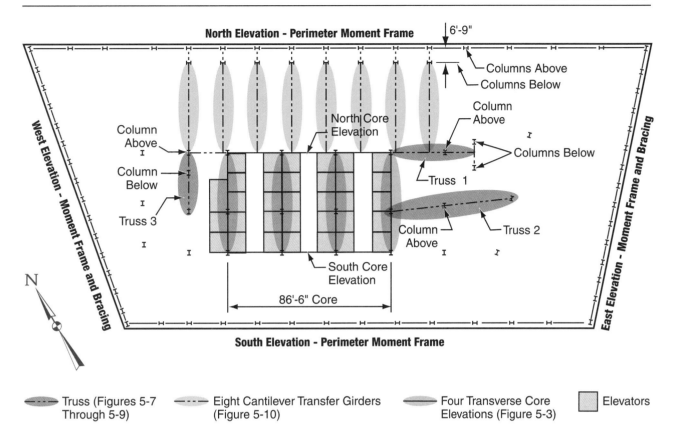

Figure 5-6 Seventh floor plan showing locations of transfer trusses and girders.

Truss 2 was a single transfer located south of Truss 1. As shown in Figure 5-8, Truss 2 transferred the column load from the 7th floor through a triangular truss into two existing columns at the 5th floor. Large gusset plates were provided at the connection between the diagonals, the columns, and the horizontal tie. The diagonals and the built-up horizontal tie were field-welded.

Truss 3 was a cantilevered two-story transfer structure in the north-south direction between the 5th and 7th floors at the western end of the core area. As shown in Figure 5-9, Truss 3 transferred the loads between columns. The upper columns carried 41 floors of load and were cantilevered to the north of the column that went from the foundation to the 7th floor.

The cantilever transfer girders, shown in Figure 5-10, spanned between the core and the north elevation at the 7th floor. There were eight transfer girders to redirect the load of the building above the 7th floor into the columns that went through the Con Ed substation. These girders cantilevered 6 feet 9 inches between the substation and the north façade of the building above. The girders extended an additional 46 feet to the core. The two transfer girders at the east end of the building were connected to Truss 1, creating a double transfer. The girders varied in depth from 9 feet at the north end, to a tapered portion in the middle, and to 4 feet 6 inches at the southern section closest to the core. Each transfer girder weighed approximately 52 tons. At the north wall, between the 7th and 5th floors, transferred columns were also part of the belt truss that circled the building as part of the lateral-load-resisting system and acted as a transfer for the columns above the shipping ramp.

CHAPTER 5: *WTC 7*

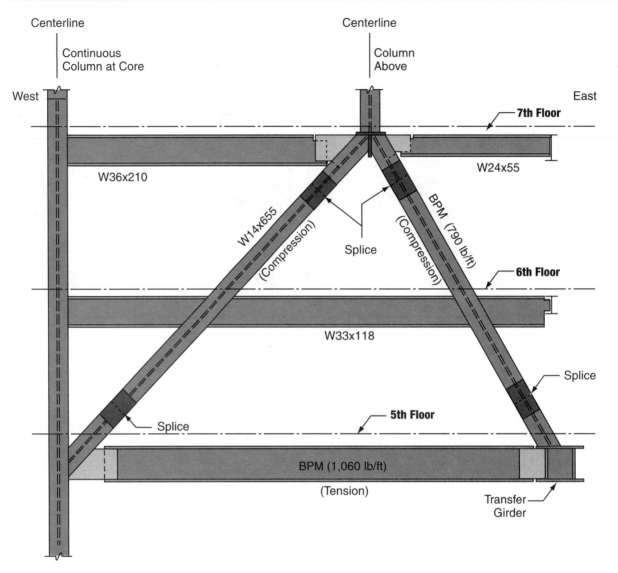

Figure 5-7   Truss 1 detail. (BPM = built-up plate member.)

5.2.4 Connections

A variety of framing connections were used. Seated beam connections were used between the exterior columns and the floor beams. Single-plate shear connections were generally used at beam-to-beam connections. Double-angle connections were provided between some beam and end-plate connections at beam-to-interior columns. Floor-framing connections used 7/8-inch-diameter ASTM A325 bolts; connections for bracing, moment frames, and column splices used 1-inch diameter ASTM A490 bolts.

Along the east and west elevations, center-to-center column spacing was typically less than 10 feet. Column trees were used at these locations. A column tree is a shop-fabricated column assembly with beam stubs shop-welded to the column flanges. The field connections were made at the end of the stubs at the center of the span between columns. One-sided lap plates were used for both flange and web connections.

Along the north and south elevations, and within the core up to the 7th floor, the spans were approximately 28 feet. At these locations, traditional moment frame construction was used. Top and bottom flange plates, as well as one-sided web shear plates, were shop-welded to column flanges. The beams were then field-bolted into the connection.

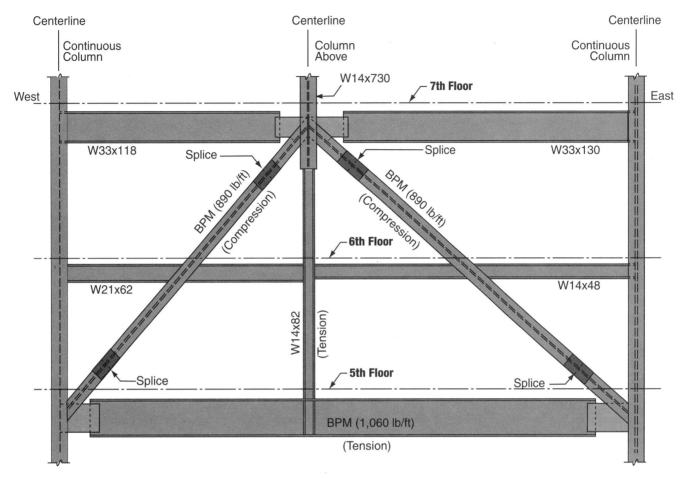

Figure 5-8 Truss 2 detail. (BPM = built-up plate member.)

The majority of column splices were bolted according to American Institute of Steel Construction (AISC) details. They were located 3 feet 6 inches above the floor and were not designed to accommodate tensile forces. Columns below the 7th floor were often "jumbo" shapes (W14x455 to W14x730) or built-up jumbo box shapes with plates up to 10 inches thick welded from flange to flange, parallel to the web, to provide the necessary section properties. For these massive columns, either the upper shaft was beveled to be field-welded or side plates were shop-welded to the lower shaft and field-welded to the upper shaft once the column was erected and plumb.

The majority of the bracing members were two channels or two T-sections connected to the structure by a welded gusset plate. A single wide flange cross-section was also used. These members were connected with web and flange plates, similar to those used in the moment frames. Some of the bracing members on the east and west sides of the building were as large as the jumbo column sections. Large connection plates were sandwiched to each side of these large braces, beams, and columns at their junctions. Bolts attached all the components to each other at these joints.

The granite façade panels were manufactured off site and were supported by individual trusses. Each panel had a single vertical/gravity connection and top and bottom lateral/wind connections to transmit these forces back to the base building. Horizontal panel adjustments could be accommodated within the panel itself. The building columns had welded angles and channels that provided horizontal and lateral support. The top of the panel was connected to the angle, and the bottom of the panel was connected to the channel. These steel-panel connections had vertically slotted holes for vertical adjustment.

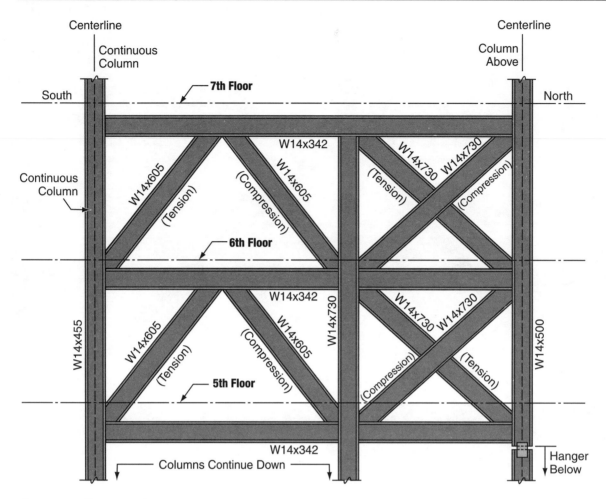

Figure 5-9   Truss 3 detail.

5.3 Fire Protection Systems

5.3.1 Egress Systems

There were two main exit stairways in WTC 7. Stairway 1 was located on the west side, and Stairway 2 was located on the east side within the central core. Both exit stairways discharged directly to the exterior at ground level and were approximately 4 feet 10 inches wide. The stairways were built of fire-rated construction using gypsum wallboard. Subsequent to the 1993 bombing incident at the WTC, battery-operated emergency lighting was provided in the stairways and photoluminescent paint was placed on the edge of the stair treads to facilitate emergency egress. In addition to the battery-powered lighting, the stairs also had emergency system lighting powered by the generators.

Twenty-eight passenger elevators and three service elevators served the various levels of WTC 7. Occupants using the elevators would typically discharge at the third level and either exit through the Lobby to bridges bringing them over to the WTC Plaza, or proceed down the escalators to grade level.

5.3.2 Detection and Alarm

Smoke detectors were located in telecommunications, electrical, and communications closets, as well as inside the HVAC system ducts, in the mechanical rooms, and in all elevator lobbies. Manual pull stations were provided at the entrances to stairways and at each of the exits. Speakers for voice evacuation

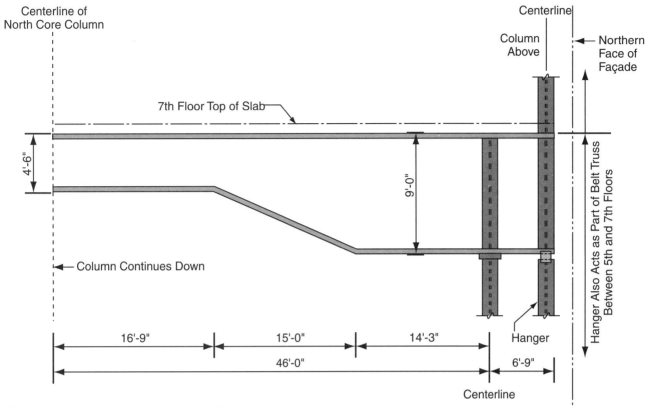

Figure 5-10 Cantilever transfer girder detail.

announcements were located throughout the building and were activated manually at the Fire Control Center (FCC). Strobes were provided and were activated automatically upon detection of smoke, water flow, or initiation of a manual pull station. Monitoring of the fire-alarm control panel for WTC 7 was provided independently at a central station. In addition to the emergency generators, the existing uninterruptible power supply (UPS) provided 4 hours of full operation for the fire-alarm system and 12 hours of standby operation. The floor contained a combination of area smoke and heat detectors.

5.3.3 Compartmentalization

Concrete floor slabs provided vertical compartmentalization to limit fire and smoke spread between floors (see Figure 5-11). Architectural drawings indicate that the space between the edge of the concrete floor slab and curtain wall, which ranged from 2 to 10 inches, was to be filled with firestopping material.

A zoned smoke control system was present in WTC 7. This system was designed to pressurize the floors above and below the floor of alarm, and exhaust the floor of alarm to limit smoke and heat spread.

The fireproofing material used to protect the structural members has been identified by Silverstein Properties as "Monokote." The Port Authority informed the BPS Team that New York City Building Code Construction Classification 1B (2-hour rating for beams, girders, trusses, and 3-hour rating for columns) was specified for WTC 7 in accordance with the architectural specifications on the construction notes drawing PA-O. According to the Port Authority, the construction notes on drawing PA-O also specified the following:

- Exterior wall columns (columns engaged in masonry walls) shall be fireproofed on the exterior side with 2-inch solid gypsum, 3-inch hollow gypsum, 2-inch concrete or spray-on fireproofing.

- Interior columns shall be fireproofed with materials and have rating conforming with Section C26-313.3 (27-269 current section).

CHAPTER 5: *WTC 7*

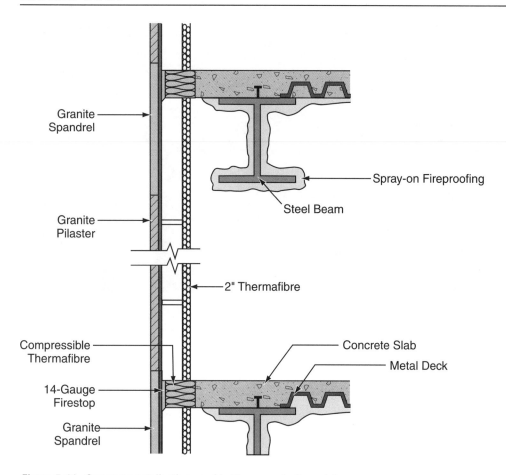

Figure 5-11 Compartmentalization provided by concrete floor slabs.

- Beams and girders shall be fireproofed with 2-inch grade Portland cement concrete, Gritcrete, or srpay-on fireproofing or other materials rendering a 2-hour fire rating.

The Port Authority stated that it believed the thickness of the spray-on fireproofing was determined by the fireproofing trade for the specific structural sections used, based on the Underwriters Laboratories formula for modifications, which were reviewed by the Architect/Engineer of Record during the shop drawing submittal. Spray-on fireproofing, as required by the code, was also listed on the drawing as an item subject to controlled inspections, in accordance with Section C26-106.3 (27-132 current section). The Architect/Engineer of Record was responsible for ensuring that the proper thickness was applied. The Port Authority had extended its fireproofing inspection program to this building.

5.3.4   Suppression Systems

The primary water supply appears to have been provided by a dedicated fire yard main that looped around most of the complex. This yard main was supplied directly from the municipal water supply. Fire department connections were located on the south and west sides.

WTC 7 was a sprinklered building. However, only the core spaces on the 5th floor were sprinkler protected, and none of the electrical equipment rooms in the building were sprinkler protected. The sprinkler protection was of "light hazard" design. The sprinkler system on most floors was a looped system fed by a riser located in Stairway 2. The loading dock was protected with a dry-pipe sprinkler system. The area of the fuel tank for OEM had a special fire detection and suppression system.

The Fire Pump Room was located on the ground floor in the southwest corner of the building and contained an automatic (as well as a manual) fire pump. There were two Fire Department of New York connections in the southwest quadrant - one on the south façade and one on the west façade.

Each stairway had standpipes in it. At each floor in each stairway, there was a 2-1/2-inch outlet with a 1-1/2-inch hose (with a 3/4-inch nozzle). In addition, the east side of each floor also had a supplemental fire hose cabinet. Primary water supply to the standpipe system came from a yard main, which was fed from the municipal water supply.

5.3.5 Power

Power to WTC 7 entered at 13,800 volts (V), was stepped down to 480/277 V by silicone oil-filled transformers in individual masonry vaults on the 5th floor, and was distributed throughout the building. On each floor, one of the 277 legs was tapped and stepped down to supply single-phase 120-V branch circuits. The main system had ground fault protection. Emergency power generators were located on various levels and provided a secondary power supply to tenants. This equipment supplied backup power for communications equipment, elevators, emergency lighting in corridors and stairwells, and fire pumps. Emergency lighting units in the exit stairways, elevator lobbies, and elevator cabs were equipped with individual backup batteries.

The tanks that provided fuel for the emergency generators were located in the building. The Silverstein and Salomon Smith Barney (SSB) fuel tanks were underground below the loading dock. The OEM tank was on the ground floor on a fire-rated steel platform within a 4-hour fire-rated enclosure. SSB had supply and return piping to the emergency generators made from a 2-1/2-inch double-wall steel pipe with a 4-inch outside diameter. The SSB fuel oil riser was single-wall pipe with a masonry shaft. Only the horizontal piping on the 5th floor was a double-wall pipe within a pipe. The pumps located at the ground floor could supply 75 gallons per minute (gpm). A 3-gpm fuel supply rate was needed for each of the nine 1,725-kilowatt (kW) generators located on the 5th floor. One gallon would be consumed and the other 2 gallons would continue to circulate through the system. The SSB fuel oil pumps were provided with UPS power supported by both base building emergency power and SSB standby power. The volume between the inner and outer pipes was designed to contain a leak from the inner pressurized pipe and direct that fuel oil to a containment vessel. Upon detection of fuel oil in the containment vessel, the fuel oil pumps automatically de-energized. The SSB fuel oil pumps and distribution piping were dedicated to the SSB generator plant. The base building life safety generators and OEM generators had their own dedicated fuel oil pumps and piping. The Silverstein generators consisted of two 900-kW units, which were also located on the 5th floor, and supplied by a 275-gallon day tank. Other characteristics of the design or controls for the fuel system for the generators are unknown.

5.4 Building Loads

The degree of impact damage to the south façade could not be documented. However, damage was evident from review of photographs and video records. The number of fires observed after the collapse of WTC 1 also makes it likely that debris impact damage occurred in a number of locations.

An array of fuels typically associated with offices was distributed throughout much of the building. In addition, WTC 7 contained 10 transformers at street level, 12 transformers on the 5th floor, and 2 dry transformers on the 7th floor. The Con Ed substation contained (outside the building footprint) eight 30-foot-wide transformers that supplied 13-kilovoltampere (kVA) power to the 6th floor of the building. Fuel oil (ranging from diesel to #4) was provided for the generators serving OEM, SSB, Silverstein Properties, and the U.S. Secret Service. Table 5.2 shows where the generators, fuel tanks, pumps, and risers were located for the various occupants. There was also a Con Ed 4-inch-diameter gas line with 0.25 pounds per square inch (psi) (low) pressure going into WTC 7 for cooking purposes. Early news reports had indicated that a high-pressure, 24-inch gas main was located in the vicinity of the building; however, this proved not to be true.

Table 5.2 WTC 7 Fuel Distribution Systems

	Storage Tanks	Pumps	Riser	Day Tank	Generators
OEM	Used Silverstein tank to fill day tanks	Ground floor; 33.3 gpm	Located in shaft in west elevator bank	275-gallon tank on 7th floor; one 6,000-gallon tank located between low-rise elevators in east elevator shaft between 2nd and 3rd floors	Three 500-kW on 7th floor on south side
Salomon Smith Barney	Two 6,000-gallon tanks under loading dock on ground level	In Fire Pump Room, west side of ground floor; 75 gpm	Located in shaft in mechanical rooms on southwest corner of building	None; pressurized recirculating loop with 2.5-inch-inside-diameter double-wall supply and return steel pipe on 5th floor	Nine 1,725-kW on 5th floor, six on north side, three in southwest corner
Silverstein Properties	Two 12,000-gallon tanks under loading dock on ground level	Between elevator shafts on west side of ground floor; 4.4 gpm	Located in shaft in west elevator bank	275-gallon tank on 5th floor	Two 900-kW on 5th floor in southwest corner
U.S. Secret Service	Used Silverstein tank	Used Silverstein pumps	Located in shaft in west elevator bank	Approximately 50- to 100-gallon tank under generator on 9th floor	9th floor
American Express	Day tank only	None	None	275-gallon tank on 8th floor on west side next to exterior wall	8th floor

As described in Section 5.6.2, the sequence of the WTC 7 collapse is consistent with an initial failure that occurred internally in the lower floors on the east side of the building. The interest in fuel oil is therefore directed at the parts of the fuel oil distribution system having the potential of supporting a fire in the lower floors on the east side of the building. The risers for the fuel distribution system were in one of the two utility shafts in the west end of the building. One exception was the American Express Corporation, which had a generator with a 275-gallon tank on the west end of the 8th floor. This tank was the sole supply for the American Express generator and was not connected to any other fuel oil source. The 275-gallon tank was filled by bringing containers of fuel oil to the tank and transferring the oil into the tank. Except for the part of the diesel oil distribution system serving the SSB generators, all of the generators were located at the west end, with relatively short horizontal distribution piping.

The SSB system involved three separate generator locations on the 5th floor: three generator sets in the southwest corner of the building, two in the northwest section, and four in the northeast section. The distribution pipe was double-wall welded black iron with leak detection between the pipes. The outer pipe was at least 4 inches in diameter and the inner pipe at least 2-1/2 inches. The pipe traversed most of the length of the 5th floor immediately north of a concrete masonry wall running most of the length of the floor in an east-west direction. At the east end of the 5th floor and to the south of the wall was a 1- to 2-story mechanical equipment room. Transfer Trusses 1 and 2 were located in this room. The east end of Truss 1 was supported by a truss element that ran perpendicular (i.e., north-south) to the main east-west portions of the truss. There was a set of double doors opening from the mechanical room to the area containing the four generator sets previously mentioned. The fuel oil distribution pipe ran above this door several feet to the north of the masonry wall. The type, quality, and hardware on the door set are unknown. The position of the door (i.e., open or closed) at the time of the incident is also unknown. Also, no information was available in regard to the size of the undercut on the door.

The fuel oil pumps were powered from the generator sets. Fuel oil would have been pumped from the tanks when the emergency power system sensed a power interruption. The pump then operated in response to the pressure difference between the supply and return, and the pump would circulate oil as long as such a difference existed. Upon sensing a power interruption, the system would automatically switch to emergency mode. This would have been done with a transfer switch that monitored the building power supply and transferred to the emergency power system if the power from the Con Ed source was interrupted. It was also possible for the transfer to be made manually. Relative to continuity of power to the building, Con Ed reported that "the feeders supplying power to WTC 7 were de-energized at 9:59 a.m." It is believed that the emergency generators came on line immediately. It is also believed that some of them may have stopped operating because of the contamination of the intake air flowing into the carburetors and radiators. Except for the SSB system, where it is understood that a UPS system provided backup power to the 75-gpm pump, the flow of oil would stop and, as soon as the day tanks were empty, the involved generator set would stop running.

The SSB generators did not use day tanks. Instead there was a pressurized loop system that served all nine generators. As long as the 75-gpm pump continued to operate, a break in the line could, under some conditions, have a full or partial break that would not cause the system to shut down and could discharge up to the 75-gpm capacity of this positive displacement pump. It is understood that the SSB pump was supplied power from both the SSB generators and from the UPS.

Engineers from the New York State Department of Environmental Conservation investigated oil contamination in the debris of WTC 7. Their principal interest was directed to the various oils involved in the Con Ed equipment. However, they reported the following findings on fuel oil: "In addition to Con Ed's oil, there was a maximum loss of 12,000 gallons of diesel from two underground storage tanks registered as 7WTC." To date, the NY State Environmental Protection Agency (EPA) and DEC have recovered approximately 20,000 gallons from the other two intact 11,600-gallon underground fuel oil storage tanks at WTC 7.

Based on the listings in Table 5.2, it is probable that the 20,000 gallons that were recovered were from the Silverstein Properties' emergency power system. The data obtained from Silverstein indicate that the pumping rate from their tanks was 4.4 gpm. If the Silverstein pump had started pumping at 10 a.m., when Con Ed shut down power to the building immediately following the collapse of WTC 2, and continued pumping until the collapse of WTC 7 at 5:20 p.m., less than 2,000 gallons would have been used. The residual 20,000 gallons found in the two 12,000-gallon tanks, therefore, can not be used as an indicator of whether or not the Silverstein generator sets were on line and running.

Similarly, the SSB pump, which had a pumping rate of 75 gpm, would have drained the two 6,000-gallon tanks serving that system in less than 3 hours. This could have accounted for the lost 12,000 gallons reported by EPA or the tanks could have been ruptured and the oil spilled into the debris pile. Again, this is not a valid indicator of whether or not the SSB generator sets came on line. The NY State EPA indicates that the SSB tanks will be pulled from the debris in the near future. This may or may

CHAPTER 5: WTC 7

not give some indication of the amount of oil still in the tanks when they were crushed. If there is evidence that the majority of diesel fuel was still in the tanks, it can be concluded that the SSB system did not discharge diesel oil as hypothesized in Section 5.6.1. Conversely, evidence that indicates that the tanks were low on oil at the time of rupture, and that they were full at the start of the September 11 incident, would lend support to the hypothesis that the SSB system was operating and pumping oil from these tanks.

Currently, there are no data available on the post-collapse condition of the OEM 6,000-gallon tank located between the 2nd and 3rd floors. The OEM system also included a 275-gallon day tank located on the 7th floor. The OEM system had a fuel supply system with the capability of transferring fuel from the Silverstein tanks to the 6,000-gallon OEM tank. The OEM generator sets were located in the southwest portion of the 7th floor. OEM also had an 11,000-gallon potable water tank on the south side of the 7th floor.

The Secret Service diesel distribution system, like the OEM system, was designed to refurbish its supply from the Silverstein tanks. This appears to have been pumped directly to a day tank having an estimated capacity of 50 to 100 gallons located near the northwest corner of the 9th floor. The generator set was also in the same location.

The 275-gallon tank associated with the American Express generator was located at the west end of the 8th floor. If full, the 275 gallons represent a potential of about 600 MegaJoules, which would be enough to cause a serious fire that could spread to other fuels, but not felt to be enough to threaten the stability of the building's structural elements.

5.5 Timeline of Events Affecting WTC 7 on September 11, 2001

The effects of the collapse of WTC 1 and WTC 2, the ensuing fires in WTC 7, and the collapse of WTC 7 are discussed below. Figure 5-12 shows the vantage points of the photographs taken illustrating these effects, as well as the extent of the debris generated by each of the collapses.

5.5.1 Collapse of WTC 2

At 9:59 a.m., WTC 2 (the south tower) collapsed. The approximate extent of its debris is shown in Figure 5-12(A). It appears that the collapse of WTC 2 did not significantly affect the roof, or the east, west, and north elevations of WTC 7. It is unknown if there was any damage to the south elevation after WTC 2 collapsed, but both the covered, tubular pedestrian bridge (see Figure 5-13) and the Plaza bridge were still standing after the collapse of WTC 2.

5.5.2 Collapse of WTC 1

At 10:29 a.m., WTC 1 (the north tower) collapsed, sending its debris into the streets below. The extent and severity of the resulting damage to WTC 7 are currently unknown. However, from photographic evidence and eyewitness accounts discussed below, it was assumed that the south side of the building was damaged to some degree and that fires in WTC 7 started at approximately this time.

Figure 5-14 is an aerial photograph that shows the debris clouds spreading around WTC 7 just after the collapse of WTC 1. Figure 5-15 is a photograph of WTC 1 debris between the west elevation of WTC 7 and the Verizon building. Figure 5-12(B) shows a plan-view diagram approximating the extent of this debris just after the collapse of WTC 1.

It does not appear that the collapse of WTC 1 affected the roof, or the east, west, and north elevations of WTC 7 in any significant way. However, there was damage to the southwest corner of WTC 7 at approximately floors 8 to 20, 24, 25, and 39 to 46, as shown in Figure 5-16, a photograph taken from West Street.

CHAPTER 5: *WTC 7*

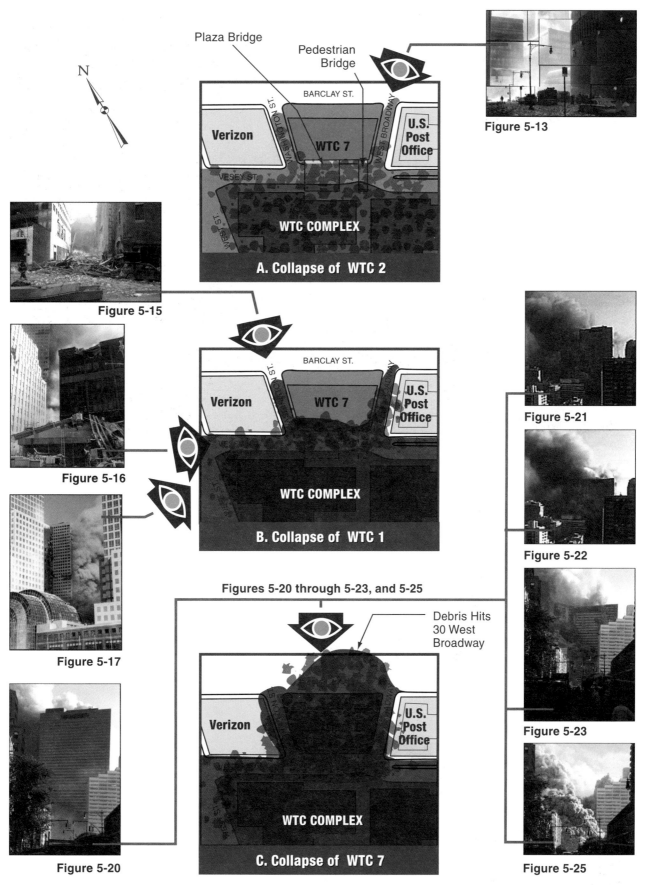

Figure 5-12 Sequence of debris generated by collapses of WTC 2, 1, and 7.

CHAPTER 5: *WTC 7*

Figure 5-13 Pedestrian bridge (bottom center) still standing after WTC 2 has collapsed, sending substantial dust and debris onto the street, but before WTC 1 (top center) has collapsed.

Figure 5-14
View from the north of the WTC 1 collapse and the spread of debris around WTC 7. Note the two mechanical penthouses of WTC 7 are intact.

CHAPTER 5: *WTC 7*

Figure 5-15 Debris from the collapse of WTC 1 located between WTC 7 (left) and the Verizon building (right).

Figure 5-16
Damage to southwest corner of WTC 7 (see box), looking from West Street.

CHAPTER 5: WTC 7

Figure 5-17, a photograph taken across from the World Financial Center (WFC), shows the west elevation and indicates damage at the southwest corner of WTC 7 at the 24th, 25th, and 39th through 46th floors.

According to the account of a firefighter who walked the 9th floor along the south side following the collapse of WTC 1, the only damage to the 9th floor façade occurred at the southwest corner. According to firefighters' eyewitness accounts from outside of the building, approximately floors 8-18 were damaged to some degree. Other eyewitness accounts relate that there was additional damage to the south elevation.

5.5.3 Fires at WTC 7

Currently, there is limited information about the ignition and development of fires at WTC 7, as well as about the specific fuels that may have been involved during the course of the fire. It is likely that fires started as a result of debris from the collapse of WTC 1.

According to fire service personnel, fires were initially seen to be present on non-contiguous floors on the south side of WTC 7 at approximately floors 6, 7, 8, 10, 11, and 19. The presence of fire and smoke on

Figure 5-17
Building damage to the southwest corner and smoke plume from south face of WTC 7, looking from the World Financial Plaza. Note damage to WFC 3 in the foreground.

lower floors is also confirmed by the early television news coverage of WTC 7, which indicated light-colored smoke rising from the lower floors of WTC 7.

Video footage indicated that the majority of the smoke appeared to be coming from the south side of the building at that time as opposed to the other sides of the building. This is corroborated by Figure 5-17, a photograph taken at 3:36 p.m. that shows the south face of WTC 7 covered with a thick cloud of smoke, and only small amounts of smoke emanating from the 27th and 28th floors of the west face of WTC 7.

News coverage after 1:30 p.m. showed light-colored smoke flowing out of openings on the upper floors of the south side of the building. Another photograph (Figure 5-18) of the skyline at 3:25 p.m., taken from the southwest, shows a large volume of dark smoke coming from all but the lowest levels of WTC 7, where white smoke is emanating. The mode of fire and smoke spread was unclear; however, it may have been propagated through interior shafts, between floors along the south façade that may have been damaged, or other internal openings, as well as the floor slab/exterior façade connections.

It appeared that water on site was limited due to a 20-inch broken water main in Vesey Street. Although WTC 7 was sprinklered, it did not appear that there would have been a sufficient quantity of water to control the growth and spread of the fires on multiple floors. In addition, the firefighters made the decision fairly early on not to attempt to fight the fires, due in part to the damage to WTC 7 from the collapsing towers. Hence, the fire progressed throughout the day fairly unimpeded by automatic or manual suppression activities.

A review of photos and videos indicates that there were limited fires on the north, east, and west faces of the building. One eyewitness who saw the building from a 30th floor apartment approximately 4 blocks

Figure 5-18 WTC 7, with a large volume of dark smoke rising from it, just visible behind WFC 1 (left). A much smaller volume of white smoke is seen rising from the base of WTC 7. Note that the lower, lighter-colored smoke (to right) is thought to be from the two collapsed towers

away to the northwest noted that fires in the building were not visible from that perspective. On some of the lower floors, where the firefighters saw fires for extended periods of time from the south side, there appeared to be walls running in an east to west direction, at least on floors 5 and 6, that would have compartmentalized the north side from the south side. There were also air plenums along the east and west walls and partially along the north walls of these floors instead of windows that may have further limited fires from extending out of these floors and, therefore, were not visible from sides other than the south.

As the day progressed, fires were observed on the east face of the 11th, 12th, and 28th floors (see Figure 5-19). The Securities and Exchange Commission occupied floors 11 through 13. Prior to collapse, fire was seen to have broken out windows on at least the north and east faces of WTC 7 on some of the lower levels.

On the north face, photographs and videos show that the fires were located on approximately the 7th, 8th, 11th, 12th, and 13th floors. American Express Bank International occupied the 7th and 8th floors. The 7th floor also held the OEM generators and day tank. Photographs of the west face show fire and smoke on the 29th and 30th floors.

Figure 5-19
Fires on the 11th and 12th floors of the east face of WTC 7.

It is important to note that floors 5 through 7 contained structural elements that were important to supporting the structure of the overall building. The 5th and 7th floors were diaphragm floors that contained transfer girders and trusses. These floors transferred loads from the upper floors to the structural members and foundation system that was built prior to the WTC 7 office tower. Fire damage in the 5th to 7th floors of the building could, therefore, have damaged essential structural elements.

With the limited information currently available, fire development in this building needs additional study. Fires were observed to be located on the lower levels for the majority of the time from the collapse of WTC 1 to the collapse of WTC 7. It appears that the sprinklers may not have been effective due to the limited water on site, and that the development of the fires was not significantly impeded by the firefighters because manual firefighting efforts were stopped fairly early in the day.

Available information indicates that fires spread horizontally and vertically throughout the building during the course of the day. The mode of spread was most likely either along the south façade that was damaged, or internally through shafts or the gap between the floor slab and the exterior wall. It is currently unclear what fuel may have been present to permit the fires to burn on these lower floors for approximately 7 hours. The change in the color and buoyancy of the smoke as the day progressed may indicate a change in the behavior of the fires. The darker color may be indicative of different fuels becoming involved, such as fuel oil, or the fire becoming ventilation limited. The increased buoyancy of the fires suggests that the heat release rate (or "fire size") may have also increased.

The mechanisms behind these apparent changes in behavior are currently unknown and therefore various scenarios need to be investigated further. These include gathering additional information regarding storage of materials on various levels, the quantity and combustibility of materials, and the presence of dense storage, including file rooms, tape vaults, etc. In addition, further analysis is needed on the specific locations of the fuel tanks, supply lines, fuel pumps, and generators to determine whether it may have been possible for a fuel line to be severed by the falling debris, allowing the pumps to run and pump fuel out of the broken pipes.

5.5.4 Sequence of WTC 7 Collapse

Approximately 7 hours after fires initiated in WTC 7, the building collapsed. The start of a timed collapse sequence was based on 17:20:33, the time registered by seismic recordings described in Table 1.1 (in Chapter 1). The time difference between each of the figures was approximated from time given on the videotape. Figures 5-20 to 5-25 illustrate the observed sequence of events related to the collapse.

~5:20:33 p.m.   WTC 7 begins to collapse. Note the two mechanical penthouses at the roof on the east and west sides in Figure 5-20.

~5:21:03 p.m.   Approximately 30 seconds later, Figure 5-21 shows the east mechanical penthouse disappearing into the building. It takes a few seconds for the east penthouse to "disappear" completely.

~5:21:08 p.m.   Approximately 5 seconds later, the west mechanical penthouse disappears (Figure 5-22) or sinks into WTC 7.

~5:21:09 p.m.   Approximately 1 or 2 seconds after the west penthouse sinks into WTC 7, the whole building starts to collapse. A north-south "kink" or fault line develops along the eastern side as the building begins to come down at what appears to be the location of the collapse initiation (see Figures 5-23 and 5-24).

~5:21:10 p.m.   WTC 7 collapses completely after burning for approximately 7 hours (Figure 5-25). The collapse appeared to initiate at the lower floors, allowing the upper portion of the structure to fall.

CHAPTER 5: WTC 7

Figure 5-20
View from the north of WTC 7 with both mechanical penthouses intact.

The debris generated by the collapse of WTC 7 spread mainly westward toward the Verizon building, and to the south. The debris significantly damaged 30 West Broadway to the north, but did not appear to have structurally damaged the Irving Trust building at 101 Barclay Street to the north or the Post Office at 90 Church Street to the east. The average debris field radius was approximately 70 feet. Figures 5-12(C) and 5-26 show an approximation of the extent of the debris after the collapse of WTC 7.

5.6 Potential Collapse Mechanism

5.6.1 Probable Collapse Initiation Events

WTC 7 collapsed approximately 7 hours after the collapse of WTC 1. Preliminary indications were that, due to lack of water, no manual firefighting actions were taken by FDNY.

Section 5.5.4 describes the sequence of the WTC 7 collapse. The described sequence is consistent with building collapse resulting from an initial (triggering) failure that occurred internally in the east portion of a lower floor in the building. There is no clear evidence of exactly where or on which floor the initiating failure occurred. Possibilities can be divided into three potential scenarios based on floor. In each

Figure 5-21
East mechanical penthouse collapsed. (From video.)

Figure 5-22
East and now west mechanical penthouses gone. (From video.)

CHAPTER 5: WTC 7

Figure 5-23
View from the north of the "kink" or fault developing in WTC 7.

case, the concern is the failure of either a truss or one or more columns in the lower floors of the east portion of the building. Each of the scenarios is a hypothesis based on the facts known and the unknown conditions that would be required for the hypothesis to be valid. The cases are presented not as conclusions, but as a basis for further investigation.

4th Floor Scenarios. The bottom cords of the transfer trusses were part of the support of the 5th floor slab and, as such, were located below the slab and above the ceiling of the 4th floor in a position exposed to fire from below. The bottom cord members were massive members weighing slightly over 1,000 pounds per foot. Such members are slow to heat up in a fire. It was reported that these bottom cords were fireproofed. The space below was the cafeteria dining room. The best information available indicates that the dining room was furnished with tables and chairs. The intensity and duration of a fire involving these furnishings would not be expected to sufficiently weaken either the trusses or the columns supporting the trusses. Member collapse as a result of a fire on the 4th floor would require either that there was significant additional fuel or that the fireproofing on the trusses or columns was defective. Fuel oil leakage from the 5th floor is also a possibility; however, no evidence of leakage paths in the east end of the second floor was reported.

CHAPTER 5: *WTC 7*

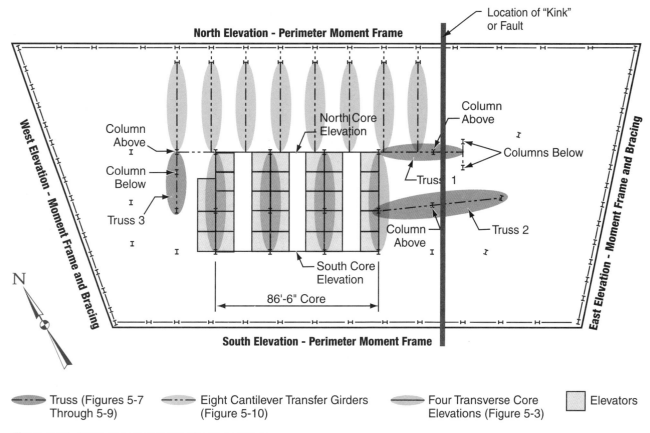

Figure 5-24 Areas of potential transfer truss failure.

Figure 5-25 Debris cloud from collapse of WTC 7.

Figure 5-26 Debris generated after collapse of WTC 7.

5th Floor Scenarios. From a structural standpoint, the most likely event would have been the collapse of Truss 1 and/or Truss 2 located in the east end of the 5th and 6th floors. These floors are believed to have contained little if any fuel other than the diesel fuel for the emergency generators, making diesel oil a potential source of fire. As noted in Section 5.4, the fuel distribution system for the emergency generators pumped oil from tanks on the lower floors to the generators through a pipe distribution system. The SSB fuel oil system was a more likely source of fire around the transfer trusses. The SSB pump is reported as a positive displacement pump having a capacity of 75 gpm at 50 psi. Fuel oil was distributed through the 5th floor in a double-wall iron pipe. A portion of the piping ran in close proximity to Truss 1. However, there is no physical, photographic, or other evidence to substantiate or refute the discharge of fuel oil from the piping system.

The following is, therefore, a hypothesis based on potential rather than demonstrated fact. Assume that the distribution piping was severed and discharged up to 75 gpm onto the 5th floor in the vicinity of Truss 1. Seventy-five gpm of diesel fuel have the potential of approximately 160 megawatts (MW) of energy. If this burning diesel fuel formed pools around Truss 1, it could have subjected members of that truss to temperatures significantly in excess of those experienced in standard fire resistance test furnaces (see Appendix A). If the supply tanks were full at the start of the discharge, there was enough fuel to sustain this flow for approximately 3 hours. If the assumed pipe rupture were incomplete and the flow less, the potential burning rate of the discharged oil would be less, but the duration would be longer. At even a 30-gpm flow rate (about 60 MW potential), the exposed members in the truss could still be subjected to high temperatures that would progressively weaken the steel. For the above reasons, it is felt that burning of discharged diesel fuel oil in a pool encompassing Truss 1 and/or Truss 2 needs to be further evaluated as a possible cause of the building collapse.

In evaluating the potential that a fire fed by fuel oil caused the collapse, it is necessary to determine whether the following events occurred:

1. The SSB generators called for fuel. This would occur once the generators came on line.

2. The pumps came on, sending fuel through the distribution piping.

3. There was a breach in the fuel distribution piping and fuel oil was discharged from the distribution system.

    Although there is no physical evidence available, this hypothesis assumes that it is possible that both the inner and outer pipes were severed, presumably by debris from the collapse of WTC 1. Depending on ventilation sources for air, this is sufficient to flashover the space along the north wall of this floor. The temperature of the fire gases would be governed to a large extent by the availability of air for combustion. The hot gases generated would be blocked from impacting Trusses 1 and 2 by the masonry wall separating the generation area from the mechanical equipment room, assuming that this wall was still intact after collapse of the tower and there were no other significant penetrations of walls.

4. The discharged fuel must be ignited.

    For diesel oil to be ignited, there must be both an ignition source and the oil must be raised to its flash point temperature of about 60 °C (140 °F). Because there were fires on other floors of WTC 7, an assumption of ignition at this level in the building is reasonable, but without proof.

5. There is sufficient air for combustion of the discharged fuel oil.

    The air required for combustion of 75-gpm (160 MW potential) diesel fuel is approximately 100,000 cubic feet per minute (cfm). If less air is available for combustion, the burning rate will decrease proportionally. As the engine generator sets come on line, automatic louvers open and 80,000 cfm are provided for each of the nine SSB engines. A portion is used as combustion air for the drive engines; the rest is for cooling, but could supply air to an accidental fire. Given open louvers and other sources for entry of air, it is, therefore, probable that a fuel oil spill fire would have found sufficient air for combustion.

6. The hot fire gases reach and heat the critical member(s).

    For this to happen, the fire must have propagated either fuel or hot gases to the members in the truss in the mechanical equipment room. If the double door to the mechanical equipment room was either open or fell from its frame at some point, or if the door was undercut, the spilled fuel oil might have flowed into the mechanical equipment room, enveloping truss members in the main (hottest) portion of the flame. Such a situation could produce an exposure possibly exceeding that in the standard furnace test producing localized heat fluxes approaching the 200 kW/m^2 used by Underwriters Laboratories to simulate a hydrocarbon pool fire, with exposure temperatures in the range of 1,200 °C (2,200 °F). If such intense exposure existed, the steel would be weakened more rapidly than normally expected. If the door was of superior construction (as with a fire door), it is unlikely that the fire would have reached the trusses in the mechanical equipment room until such time that the door failed.

CHAPTER 5: WTC 7

A further hypothesis that would help explain the long time lapse between the collapse of WTC 1 and the collapse of WTC 7 would be that the masonry wall and door resisted the fire for a number of hours, but eventually failed. The new opening then allowed the fire (still supplied with a continuous discharge of fuel oil) to flow into the mechanical equipment room, envelope elements of the fireproofed trusses, and eventually cause a buckling collapse of one or both of them. For the fire to last long enough for this to occur, the flow rate would have to be around 30 gpm. At a rate of 30 gpm, the fuel would last for about 7 hours and would produce a fire of about 60 MW. The possibility that such a scenario could occur would be dependent on the specific construction details of the wall, the door, and the fireproofing on the truss.

Another hypothesis that has been advanced is that the pipe was penetrated by debris at a point near the southwest corner where there was more damage caused by debris from the collapse of the towers. This would have resulted in fuel oil spilling onto the 5th floor, but not being immediately ignited. However, a major portion of the 12,000 gallons in the SSB tanks would pump out onto the 5th floor, forming a large pool. At some point, this would have ignited and produced the required fire. This hypothesis has the advantage of assuming a pipe break in the area most severely impacted by the tower debris and accounts for the long delay from the initial incident to the collapse of WTC 7. The principal challenge is that such a fire would have more severely exposed Truss 3. If Truss 3 had been the point of collapse initiation, it is not expected that the first apparent sign of collapse would be the subsidence of the east penthouse.

Evaluation of fires on the 3rd to 6th floors is complicated by the fact that these floors were windowless with louvers, generally in a plenum space separating any direct line of sight between the open floor space and the louvers. None of the photographic records found so far show fires on these floors.

Further investigation is required to determine whether the preceding scenarios did or could have actually occurred.

Other Involved Floors Scenarios. Fire was known to have occurred on other floors. If a fire on one of these floors involved a large concentration of combustible material encasing several columns in the east portion of the floor, it might have been of sufficient severity to cause the structural members to weaken. Such fuel concentrations might have been computer media vaults, archives and records storage, stock or storage rooms, or other collections. It is possible that the failure of at least two or possibly more columns on the same floor would have been enough to cause collapse.

5.6.2 Probable Collapse Sequence

The collapse of WTC 7 appears to have initiated on the east side of the building on the interior, as indicated by the disappearance of the east penthouse into the building. This was followed by the disappearance of the west penthouse, and the development of a fault or "kink" on the east half of WTC 7 (see Figures 5-23 and 5-24). The collapse then began at the lower floor levels, and the building completely collapsed to the ground. From this sequence, it appears that the collapse initiated at the lower levels on the inside and progressed up, as seen by the extension of the fault from the lower levels to the top.

During the course of the day, fires may have exposed various structural elements to high temperatures for a sufficient period of time to reduce their strength to the point of causing collapse. The structural elements most likely to have initiated the observed collapse are the transfer trusses between floors 5 to 7, located on lower floors under the east mechanical penthouse close to the fault/kink location.

If the collapse initiated at these transfer trusses, this would explain why the building imploded, producing a limited debris field as the exterior walls were pulled downward. The collapse may have then spread to the west. The building at this point may have had extensive interior structural failures that then led to the collapse of the overall building. The cantilever transfer girders along the north elevation, the strong diaphragms at the 5th and 7th floors, and the seat connections between the beams and columns at the building perimeter may have become overloaded after the collapse of the transfer trusses and caused the interior collapse to propagate to the whole floor and to the exterior frame. The structural system between floors 5 and 7 appears to be critical to the structural performance of the entire building.

An alternative scenario was considered in which the collapse started at horizontal or inclined members. The horizontal members include truss tension ties and the transfer girder of the T-1 truss at the east side of the 5th floor. Inclined members spanned between the 5th and 7th floors and were located in a two-story open mechanical room. The horizontal haunched back span of the eastern cantilever transfer girders, located roughly along the kink, rested on a horizontal girder at the 7th floor supported by the T-1 transfer truss. Even if the cantilever transfer girder had initiated the collapse sequence, the back span failure would most likely have not caused the observed submergence of the east mechanical penthouse.

The collapse of WTC 7 was different from that of WTC 1 and WTC 2, which showered debris in a wide radius as their frames essentially "peeled" outward. The collapse of WTC 7 had a small debris field as the façade was pulled downward, suggesting an internal failure and implosion.

To confirm proposed failure mechanisms, structural analysis and fire modeling of fuels and anticipated temperatures and durations will need to be performed. Further study of the interaction of the fire and steel, particularly on the lower levels (i.e., 1st–12th floors) should be undertaken to determine specific fuel loads, location, potential for impact from falling debris, etc. Further research is needed into location of storage and file room combustible materials and fuel lines, and the probability of pumps feeding fuel to severed lines.

5.7 Observations and Findings

This office building was built over an electrical substation and a power plant, comparable in size to that operated by a small commercial utility. It also stored a significant amount of diesel oil and had a structural system with numerous horizontal transfers for gravity and lateral loads.

The loss of the east penthouse on the videotape suggests that the collapse event was initiated by the loss of structural integrity in one of the transfer systems. Loss of structural integrity was likely a result of weakening caused by fires on the 5th to 7th floors. The specifics of the fires in WTC 7 and how they caused the building to collapse remain unknown at this time. Although the total diesel fuel on the premises contained massive potential energy, the best hypothesis has only a low probability of occurrence. Further research, investigation, and analyses are needed to resolve this issue.

The collapse of WTC 7 was different from that of WTC 1 and WTC 2. The towers showered debris in a wide radius as their external frames essentially "peeled" outward and fell from the top to the bottom. In contrast, the collapse of WTC 7 had a relatively small debris field because the façade came straight down, suggesting an internal collapse. Review of video footage indicates that the collapse began at the lower floors on the east side. Studies of WTC 7 indicate that the collapse began in the lower stories, either through failure of major load transfer members located above an electrical substation structure or in columns in the stories above the transfer structure. Loss of strength due to the transfer trusses could explain why the building imploded, with collapse initiating at an interior location. The collapse may have then spread to the west, causing interior members to continue collapsing. The building at this point may have had extensive interior structural failures that then led to the collapse of the overall building, including the cantilever transfer girders along the north elevation, the strong diaphragms at the 5th and 7th floors, and the seat connections between the interior beams and columns at the building perimeter.

5.8 Recommendations

Certain issues should be explored before final conclusions are reached and additional studies of the performance of WTC 7, and related building performance issues should be conducted. These include the following:

- Additional data should be collected to confirm the extent of the damage to the south face of the building caused by falling debris.

- Determination of the specific fuel loads, especially at the lower levels, is important to identify possible fuel supplied to sustain the fires for a substantial duration. Areas of interest include storage rooms, file rooms, spaces with high-density combustible materials, and locations of fuel lines. The control and operation of the emergency power system, including generators and storage tanks, needs to be thoroughly understood. Specifically, the ability of the diesel fuel pumps to continue to operate and send fuel to the upper floors after a fuel line is severed should be confirmed.

- Modeling and analysis of the interaction between the fires and structural members are important. Specifically, the anticipated temperatures and duration of the fires and the effects of the fires on the structure need to be examined, with an emphasis on the behavior of transfer systems and their connections.

- Suggested mechanisms for a progressive collapse should be studied and confirmed. How the collapse of an unknown number of gravity columns brought down the whole building must be explained.

- The role of the axial capacity between the beam-column connection and the relatively strong structural diaphragms may have had in the progressive collapse should be explained.

- The level of fire resistance and the ratio of capacity-to-demand required for structural members and connections deemed to be critical to the performance of the building should be studied. The collapse of some structural members and connections may be more detrimental to the overall performance of the building than other structural members. The adequacy of current design provisions for members whose failure could result in large-scale collapse should also be studied.

5.9 References

Davidowitz, David (Consolidated Edison). 2002. Personal communication on the continuity of power to WTC 7. April.

Flack and Kurtz, Inc. 2002. Oral communication providing engineering explanation of the emergency generators and related diesel oil tanks and distribution systems. April.

Lombardi, Francis J. (Port Authority of New York and New Jersey). 2002. Letter concerning WTC 7 fireproofing. April 25.

Odermatt, John T. (New York City Office of Emergency Management). 2002. Letter regarding OEM tanks at WTC 7.

Rommel, Jennifer (New York State Department of Environmental Conservation). 2002. Oral communication regarding a November 12, 2001, letter about diesel oil recovery and spillage. April.

Salvarinas, John J. 1986. "Seven World Trade Center, New York, Fabrication and Construction Aspects," *Canadian Structural Engineering Conference.*

Silverstein Properties. 2002. Annotated floor plans and riser diagrams of the emergency generators and related diesel oil tanks and distribution systems. March.

Robert Smilowitz
Adam Hapij
Jeffrey Smilow

6 *Bankers Trust Building*

6.1 Introduction

The Bankers Trust building at 130 Liberty Street, also referred to as the Deutsche Bank building, withstood the impact of one or more pieces of column-tree debris raining down from the collapsing south tower (WTC 2). Although the debris sliced through the exterior façade, fracturing spandrel beam connections and exterior columns for a height of approximately 15 stories, the building sustained only localized damage in the immediate path of the debris from WTC 2 (hereafter referred to as the impact debris) (Figures 6-1 and 6-2). There were no fires in this building. The ability of this building to sustain significant structural damage yet arrest the progression of collapse is worthy of thorough study. Unlike WTC 1, 2, and 7, which collapsed completely, the Bankers Trust building provided an opportunity to analyze a structure that suffered a moderate level of damage, to explain the structural behavior, and to verify the analytical methods used. The following sections describe the building structure, the extent of damage, and the computational methods that were used to analyze the structure.

6.2 Building Description

The Bankers Trust building is a steel-frame commercial office structure, designed and constructed circa 1971. Bankers Trust was designed by Shreve, Lamb & Harmon Associates P. C. Architects; Peterson and Brickbauer Associated Architects; the Office of James Rudderman Structural Engineers, and Jaros Baum and Bolles Mechanical and Electrical Engineers. The building measures 560 feet in height with 40 stories above grade and 2 below. It is located directly across Liberty Street from the former site of WTC 2, about 600 feet due south of the southeast corner of WTC 2.

Figure 6-1
North face of Bankers Trust building with impact damage between floors 8 and 23.

Figure 6-2
Closeup of area of partial collapse. Note debris accumulated at the bottom of the damage area, resting on the 8th floor. Area of initial impact is not shown in this photo.

The floor numbering used in the building elevator system and referred to in this report omits the 13th floor and includes a mezzanine between the 5th and 6th floors.

Above the second floor level, the building is essentially square in plan shape, with overall dimensions (centerline of exterior column lines) of 183 feet square. At the perimeter of the building, layout of columns in both the north-south and east-west directions consists of an exterior bay on each end that is 26 feet 3 inches wide and five interior bays that are 26 feet 0 inches wide. Interior column spaces vary slightly from this to accommodate the central elevator core (see Figure 6-3).

Girders and spandrel beams are typically deep wide-flange shapes, including W24, W27, W30, and W36 sections, but also including occasional built-up sections termed "wind girders." Girders and spandrel beams are moment-connected to columns at each intersection and at both axes of the columns. The girders' moment connections to column flanges were composed of top and bottom plates fillet-welded to the beam flanges and full-penetration welded to the column flange. These connections were designed for wind moment only, not the flexural capacity of the members, and are considered to be fully restrained, partial-strength connections. Lateral drift (stiffness) due to wind loads usually controls the design for moment frames. The beam web was connected to the column via a shear plate, fillet-welded to the column and bolted to the beam web. Girders that were connected to the column web utilized top plates fillet-welded to the beam top flange and full-penetration weld to the column web; there was no connection to the inside face of the column flanges. The bottom flange of the girder was connected to an extended stiffened seat that had its seat plate full-penetration-welded to the column web. The beam web had no connection to the column because the seated connection provided the necessary shear transfer. Girder-to-column shear connections utilize A325 high-

CHAPTER 6: *Bankers Trust Building*

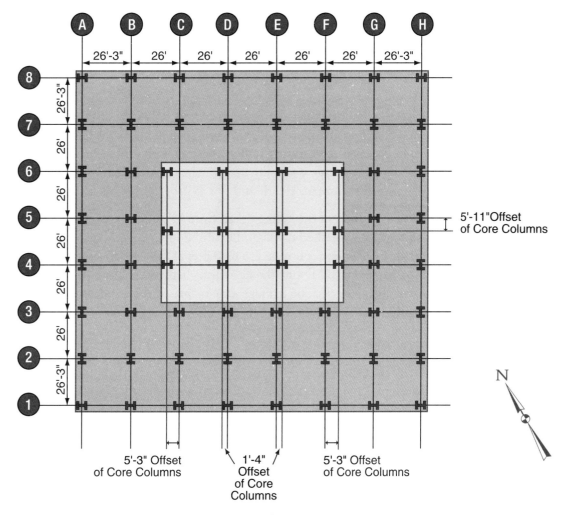

Figure 6-3   Floor plan above the 2nd level (ground floor extension not shown).

strength bolts. Steel grade is ASTM A572 grade 50 except for the wind girders, which are A36. The columns are generally A36, except that some A588 steel is used at lower-level columns.

Floor beams are typically non-composite and consist of rolled W14, W16, and W18 shapes spanning north to south between girders, typically spaced at 8 feet 8 inches. The floor beams are fastened to the girders via shear connections and typically utilize a partial end plate welded to the floor beam and bolted to the girder with A307 bearing bolts. The floor itself typically consists of 1-1/2- inch metal deck supporting a 2-1/2-inch lightweight concrete slab. Floor beams are typically depressed 2 1/2 inches relative to the girders. Typical floor height (above the 5th floor) is 12 feet 2 inches. All beam-to-beam connection bolts appear to be A307. To accommodate this limited height, limited depth members with Grade 50 (50 ksi yield strength) steel were used.

A mechanical floor is present at the 5th level, and this story is taller, at 20 feet 8 inches. The stories below the 5th level are also taller than typical, with heights varying from 16 feet to 20 feet. To laterally stiffen the frame in these tall stories, a system of diagonal bracing is provided at the elevator core.

At the first story, the building extended outward to the north, with a large canopy structure. The canopy structure extended several column lines to the north of the main building line. A full basement underlies the entire structure.

CHAPTER 6: Bankers Trust Building

6.3 Structural Damage Description

Debris from WTC 2 fell along the north side of the building. This debris completely crushed the single-story extension of the building, north of line 8, and collapsed it into the basement in this area. A column section from WTC 2 was embedded in the north edge of the floor slab of the 29th floor. It also appears that one section, or perhaps several sections, of exterior column trees from the south wall of WTC 2 plunged through the north wall of the building just above the 23rd floor. This impact area is illustrated in Figure 6-4. The zone of structural damage remained confined to one structural bay for several floors immediately below the point of impact before spreading to two and sometimes three bays in the floors below. However, although the pattern of damage was influenced by the structural response, most of the damage can be attributed to the path of the impact debris and not to progressive collapse.

It appears that the direction of motion of the falling debris from WTC 2 was steeply angled down and to the east-southeast. As the falling debris smashed through the 23rd floor spandrel between column lines C and D, it and the debris it created at each floor continued to dive deeper into the building, causing more structural damage. Between the 19th and 22nd floors, the floor areas between column lines 7 and 8, and C and D were damaged or destroyed (Figure 6-5). The damage zone increased to include portions of the area between column lines D and E from the 18th floor down to the 9th floor (Figure 6-5).

Column splices are typically located at every second floor and are composed of thin splice plates bolted to the column flanges. Large axial tension loads were probably not a design condition for the column splices, and the splice plates appear to be minimal in thickness, offering little resistance to separation. The D-8 column splices at the 18th and 16th floors appear to have been overloaded, leaving a section of column suspended from the spandrels at floors 16 through 18. The column below this level and down to the 8th

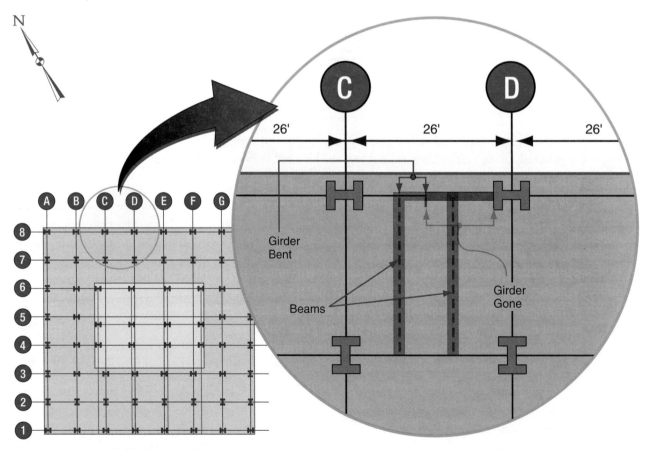

Figure 6-4 Area of initial impact of debris at the 23rd floor.

CHAPTER 6: *Bankers Trust Building*

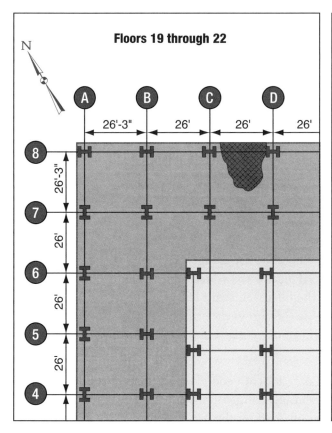
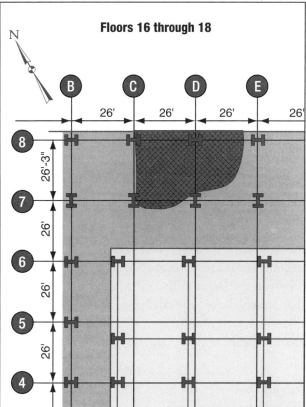
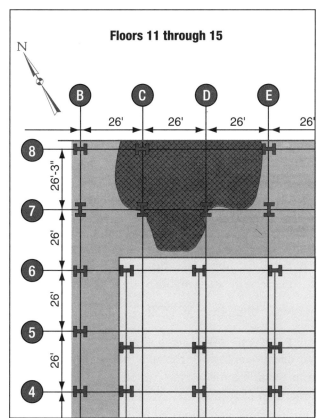
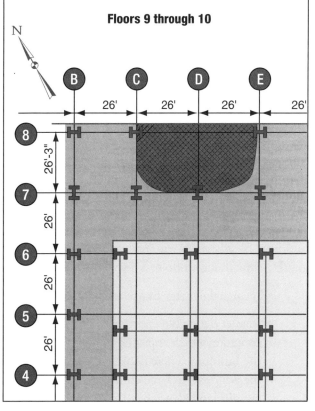

Figure 6-5   Approximate zones of damage – 19th through 22nd floors, 16th through 18th floors, 11th through 15th floors, and 9th through 10th floors.

FEDERAL EMERGENCY MANAGEMENT AGENCY

floor was either ejected from the building or folded into the general debris during the partial collapse. In the area between the 8th and 16th floors, the damage area generally increased as the partial collapse progressed downward to include, and sometimes exceed, portions of the two bays bounded by columns C-8 to E-8 and C-7 to E-7, with column D-8 missing in the area of the 12th floor.

At the point of impact and at the floors immediately below, it appears the impact debris sliced through the spandrels. However, at lower elevations, the spandrel beams were not fractured and separation occurred at the spandrel-to-column connections. In the typical failure mode of the girder-to-column-flange connection, the weld and heat-affected zone of the top flange plate pulled out of the column and left a crater in the flange. The bottom flange plate was overloaded in bending at the column face without creating a crater in the column flange. The beam shear connection was typically left in place. The shear failure occurred mostly in the beam web and, in some cases, through the bolt line in the shear connection plate.

Girders that were connected to the column web had their top flange plate overloaded in tension at the weld zone interface. Normally, a crater was created where the top flange plate was welded to the column web. The bottom flange sat on a stiffened seat. Typically, the seated connection was left in place; however, the beam pulled off of the seat, separating at the fillet welds and leaving behind the fillet welds that attached the beam flange to the seat top plate. Figure 6-6 shows the stiffened seat at the weak axis side of the column. Figure 6-7 shows that the shear connection remains on the flange side, but only a piece of beam remains hanging on the web side.

Floor-beam-to-girder connections typically used a partial-depth end plate welded to the floor beam web and bolted to the girder with A307 bearing bolts. Typically, failure occurred in one of two modes, either the bolts were overloaded in tension and the connection pulled off, or the partial end plate sheared at or near the weld line. At most of the connections, it appears there was some amount of bolt failure. Figure 6-8 shows a shear connection with half of the end plate remaining in place.

As the spandrel connections failed and the floor slabs collapsed, a portion of the rubble accumulated into a two-story pile while the remainder fell out of the building onto the low-rise roof on the north side of the building. The impact debris finally came to rest on the two-story-deep pile of debris that was on the 8th floor. Figures 6-9 and 6-10 illustrate the damage. A major component of WTC 2, approximately 40 feet tall by 30 feet wide, remained lodged in the debris pile and was clearly visible as it hung from the face of the building. Yet despite this weight, the floor supporting the debris deflected a maximum of 6 inches. Figure 6-11 shows the extent of debris in the lobby of the building.

6.4 Architectural Damage Description

In addition to the destruction of the canopy structure north of column line 8, and the collapse of the floor areas between the 8th and 23rd floors, there was general damage to the entire north façade of the building. Nearly every window was broken on the western half of the north face, between column lines B and E below the 23rd floor. This window breakage would appear to be attributable to the following causes:

- Localized damage in the areas impacted by column trees falling from WTC 2.

- Smaller debris blown from WTC 1 and WTC 2. In particular, several small chunks of lightweight concrete, which appeared to be from WTC 2 floor slabs, were thrown through the north windows of the building. These debris items ranged in size from small fragments that caused bullet-size holes in the windows to large chunks with a maximum dimension of approximately 12 inches. Many of these chunks landed as far as 15 feet from the exterior building line and appeared to be traveling almost horizontally when they penetrated the building façade.

CHAPTER 6: *Bankers Trust Building*

Figure 6-6   Moment-connected beams to columns.

Figure 6-7   Column with the remains of two moment connections.

CHAPTER 6: *Bankers Trust Building*

Figure 6-8
Failed shear connection of beam web to column web.

Figure 6-9
Suspended column D-8 at the 15th floor. Note separation at column splices.

Figure 6-10 Area of collapsed floor slab in bays between C-8, E-8, C-7, and E-7, from the 15th floor.

Figure 6-11 Bankers Trust lobby (note debris has been swept into piles).

After the exterior glazing was penetrated by the debris, the dust cloud resulting from the collapse of the towers deposited a layer of dust an inch or more thick throughout the northern part of the building. The 2nd floor lobby area had extensive broken glass, general debris, and dust (Figure 6-11). Figure 6-12 shows a typical office near the collapsed area at the 8th floor.

Although fire sprinkler piping was damaged in the collapse area, causing water to flow on the floor, in general, sprinkler piping throughout other portions of the building remained intact and the building was basically dry. Water pressure in the domestic system was available in upper floors. Ceiling systems generally remained intact, except at the collapsed areas of the building.

6.5 Fireproofing

The structural steel sections were fireproofed with a spray-applied non-asbestos fireproofing material. The thickness on the beam flanges was observed to be on the order of 1/2 inch thick. Many of the rolled steel shapes appeared to be almost completely bare of fireproofing where directly impacted by debris; the remainder of the fireproofing appeared intact even in the damaged areas. Because fires were not ignited in combination with this structural damage, the damaged fireproofing did not affect the performance of the building.

6.6 Overall Assessment

Except for the canopy structure on the north side of the building, which was crushed by falling debris, the building withstood the debris impact well. Excluding the framing and supported floors in the immediate zone of impact and several floors immediately below this area, the structure remained in good condition and serviceable. Repair of the structure should be feasible.

Figure 6-12
Office at the north side of the 8th floor.

6.7 Analysis

A 3-D model was developed to better understand the performance of the building in response to impact from debris and to identify specific design features that contributed to this performance. The model was developed based on information obtained from the following sources: structural drawings (Rudderman 1971–1975), *Draft Structural Engineering Evaluation for the New York Department of Design and Construction (*Nordenson et al. 2001*)*, and personal accounts (Smilow 2001).

All major structural entities were simulated within the ANSYS model: all major framing, intermediate framing in regions of damage, steel decking/slab, columns, and vertical bracing. Given the level of modeling detail required to simulate the behavior of the structure in the regions of damage, three levels of refinement were incorporated within the model to use analytical resources more efficiently:

Level 1: No visible structural damage

Level 2: Local structural damage

Level 3: Partial or imminent collapse

Because there are no eyewitness accounts available that can clarify the order of collapse of portions of the structure, two types of analyses were considered.

Nominally, the final state of the structure could be analyzed to determine the current state of stress redistribution within existing structural members. This analysis clarifies why the damage was arrested.

A more complex analysis could be performed, namely an analysis tracking the partial progression of damage. This type of analysis consists of parametric studies performed to determine the velocity of the impact debris that was required to:

(a) collapse the region of slab between column lines C and D at the 16th through 21st floors;

(b) remove the exterior column D-8 below the 16th floor; and

(c) extend the region of damaged slabs to the two bays bounded by column lines C and E.

These calculations are based on engineering principles and methods. Although they attempt to quantify the extent of the damaged regions, they provide little intrinsic value to the understanding of the building performance. It is highly unlikely the calculations would indicate a unique sequence of events resulting in the observed collapse patterns. Analyses of the surviving structure are much more informative and indicate the redundancy of the steel moment-frame system. These calculations indicate why the collapse was arrested at the 8th floor and why the region of slab loss did not extend beyond the two exterior bays on either side of column line D. Calculations demonstrating the redistribution of load could be performed statically, allowing for inelastic deformations.

Two different types of girder-column moment connections failed as a result of the debris impact, and the performance of these connections had a major effect on the extent of collapse. These two connections were analyzed to better understand their capacity to resist extreme loading. Although the design drawings show some connection information and the observations of the damaged building (Smilow 2001) identified some critical dimensions, there was not enough information to develop detailed finite element models. Therefore, parametric evaluations based on engineering judgment were required to identify the trends in the reserve capacities of these critical connections.

At the time this report was published, a parametric study had been completed to determine the behavior of typical moment connection details in preparation for a subsequent suite of full 3-D frame parametric studies. This suite of calculations was not completed in time for inclusion in this report.

6.7.1 Key Assumptions

Key assumptions were made in the modeling of the building. These assumptions are based on engineering judgment and basic principles of physics:

1. Boundary Conditions:

    a. The structural model was fixed at the base of Concourse Level A.

    b. Vertical rollers were placed at frame 4; given the localization of damage, approximations can be made to model half of the building and restrain the structure a single bay beyond the region of spot damage.

    c. The two-story canopy, north of frame line 8, was excluded from the model; because the structure (frame lines 8 through 10) suffered heavy damage, the collapsed region would not influence the response of the structure in the floors above.

2. Static Nonlinear Analysis:

    a. The nonlinear analysis was performed accounting for large-deflection geometric nonlinearities.

    b. The inelastic response of connections was simulated via nonlinear springs and localized inelastic material properties.

3. Multi-phase Loading:

    a. Application of the gravitational loading.

    b. Removal of missing structure; damaged and missing members were removed in a final-state analysis, and selected members were removed sequentially in an analysis tracking the partial progression of damage.

6.7.2 Model Refinement

Three levels of refinement were incorporated into the model to make more efficient use of the analytical resource for the regions of structural damage. These levels of refinement were developed based on the classification of damage patterns depicted in the draft evaluation prepared by Nordenson (2001) and supplemented by observations of structural damage (Smilow 2001).

In refinement level 1, all major framing was modeled explicitly. All moment and shear connections are assumed to be elastic within regions modeled with this refinement. The steel deck/slab system in conjunction with intermediate framing was modeled using orthotropic plates.

In refinement level 2, all major framing (spandrel beams) was modeled explicitly. Intermediate framing (beams) was not incorporated in the orthotropic plate definition; it was modeled explicitly. All moment connections are modeled explicitly, simulating all spandrel/column connections. All shear connections are assumed to be elastic within regions modeled with this refinement. The steel deck/slab system is not represented explicitly.

In refinement level 3, all major framing (spandrel beams) and intermediate framing (beams) were modeled explicitly. All moment connections and shear connections were modeled explicitly, simulating all spandrel/column connections and beam/spandrel connections, respectively. The steel deck/slab system was not represented explicitly.

Plate elements and beam elements were refined systematically to obtain key output data in regions of heavy damage with correspondingly less refinement farther away from heavy damage areas. In regions modeled using refinement level 3, a minimum of four beam elements were used between columns. The plate elements resting on the beam elements were meshed to match the resolution of the beam elements within this region of refinement. In regions modeled using refinement level 2, a minimum of two beam elements were used between columns.

The plate elements resting on the beam elements were also meshed to match the resolution of the beam elements within this region of refinement. In regions modeled using refinement level 1, single elements were used between columns. This region was modeled to simulate the behavior of regions that were not subjected to structural damage and, because no damage patterns were anticipated, the plate elements in this region of refinement were meshed with minimum discretization.

6.7.3 Simulation of Nonlinear Behavior

The ANSYS model simulates the inelastic response of the structure system by means of nonlinear springs. All moment connections and shear connections in regions of heavy damage (refinement level 3) were modeled using nonlinear rotational and translational springs.

All explicit shear connections were modeled using nonlinear vertical springs. The remaining translational degrees of freedom were assumed to be rigidly constrained. All column splices were assumed to provide continuity based on observations of the structure (Smilow 2001). All explicit moment connections were modeled using nonlinear spring elements. The properties of these nonlinear elements were determined from three-dimensional quasi-static analyses of representative connections, whose response produces a moment rotation relationship. The inelastic response of the connection was simulated by using a elasto-plastic model for the girder, plating, and weld material.

Contrary to the structural drawings, inspection of the floor structure revealed that the steel deck/slab system was not explicitly attached to the supporting spandrel beam elements; it was resting on the underlying spandrel and beam system. This structural assembly was simulated by rigidly constraining the vertical degrees of freedom of the plate elements to the underlying spandrel/beam system. This finite element construct accounted for the transmittal of weight of the deck/slab system, along with additional amounts of reported debris, to the supporting beam elements.

6.7.4 Connection Details

The behavior of a typical fully rigid, partial strength wind-moment connection about the strong axis of the column was studied. The connection of the W18x50 girder to the W14x426 column between girder line 7-8 at frame line D on the 14th floor was modeled as a representative connection. The top and bottom moment plates (estimated as 5/8 inch x 6 inches x 24 inches and 3/8 inch x 10-1/2 inches x 24 inches, respectively) were welded to girder flanges with a 1/4-inch weld. The shear plate (estimated as 5/16 inch x 3 inches x 12 inches) was bolted to each girder web with four 7/8-inch-diameter bolts. Although the design wind moment was estimated to be 2,930 kip-in, the connection capacity was estimated to be 10,800 kip-in.

Similarly, the behavior of a typical fully rigid, partial-strength wind-moment connection about the weak axis of the column was studied. The connection of the W24x68 girder to the W14x426 column between girder line C-D at frame line 7 on the 15th floor was modeled as a representative connection. The top and bottom moment plates were estimated as 3/8 inch x 12 inches x 14 inches with a 1/4-inch weld, and the shear in the connection was resisted by a seat, estimated as 1/2 inch x 5 inches x 12 inches, stiffened with a 3/8-inch x 8-inch seat plate. Although the design wind moment was estimated to be 2,830 kip-in, the connection capacity was estimated to be 7,500 kip-in, thus confirming the frame design was governed by stiffness and not strength.

Both connections were modeled in three dimensions in ANSYS, using shell, beam, and continuum elements. All weld material was simulated with a bilinear kinematic hardening material with brittle fracture capabilities at a specified ultimate strain. Figures 6-13 and 6-14 illustrate the details of the finite element models of the two different connection details.

6.7.5 Connection Behavior

Both models were subjected to numerous load combinations to determine the overall behavior of the connection. The weld material was assumed to have a nominal yield strength of 50 ksi. Each model was then subjected to a monotonically increasing moment about the transverse axis of beam bending (M_Y) and the principal strains in the welds were evaluated at the end of each load increment. If the strains in any of the weld elements exceeded the specified ultimate strain, the weld element was considered to have fractured and the modulus of elasticity was reduced by several orders of magnitude. Because the ultimate strain in the weld corresponding to fracture is an unknown quantity, several values were assumed in order to determine the connection behavior. Values of 0.5 percent, 1.0 percent, 10 percent, and 20 percent strain were assumed, and the moment curvature relations for the connection were developed. Based on these calculations, it was observed that the welds fracture before plastic hinges occur when the ultimate strain in the welds is assumed to be less than 1 percent. Furthermore, the connection was observed to degrade very quickly with the onset of weld fracture. The first onset of yielding in the welds was observed at a M_Y value of 1,000 kip-in and 1,400 kip-in, for the shear plate and seat connections, respectively. In the absence of wind moments, the connections were found to be able to support a considerable increase in gravity loads over their dead and live load design values. However, the connection offered little resistance to torsional loads and a significant reduction in capacity of the connection with respect to out of plane bending.

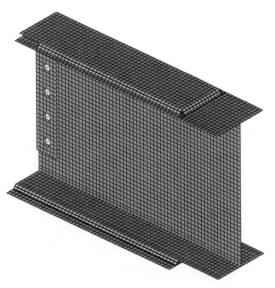

Figure 6-13
3-D ANSYS model of flange and shear plate moment connection.

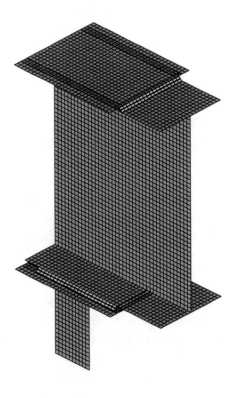

Figure 6-14
3-D ANSYS model of flange and seat moment connection.

The computed results show a sensitivity of the moment curvature relations to the ultimate strain of the weld material and the out-of-plane moments that may have been applied to the connections. The connection sensitivity to out-of-plane and twisting moments significantly influenced the capacity of the connection to resist abnormal loading. The significant reserve capacity of these connections to gravity loads, over an order of magnitude by some estimates, are quickly eroded when the connection is subjected to out-of-plane bending. Therefore, as members were twisted by the collapse of adjacent bays, the connections were less able to withstand the weight of the accumulated debris. This phenomenon may explain why many connections failed and may also explain the sequence of weld fracture. This in turn may have influenced the modes of failure in different connections.

6.8 Observations and Findings

An evaluation of the damage patterns revealed several interesting interpretations. The spandrels were sheared by the impactor, between column lines C and D, from the 23rd to the 19th floors. The D-8 column splices failed at the 18th floor and at the 16th floor, but there are no clues to indicate why column splice tension overload occurred at this location. However, unlike the spandrels above, the girder-column connections at column lines C and D failed. Although severed from the column above and below, column D-8 remained suspended from the girders spanning between column lines E and D. These girders developed large vertical and lateral deformations (twisting). The twisting and bending of these girders may have extended the zone of collapse to bays bounded by column lines C and E. If the column splices had not failed at the 16th and 18th floors, it is possible the extent of collapse may have been limited to the single bay in the path of the impactor. This enlarged zone of damage continued until the collapse was arrested on the 8th floor. It is unlikely that dynamic effects caused the damage to column D-8 below the 16th floor; otherwise, the collapse should have progressed all the way to the ground. It is possible that the column splice failures and the resulting large deformations (twisting) of the spandrels caused the remaining portion of column D-8 to lose lateral bracing, and the collapse was not arrested until the energy of the impactor and debris pile was sufficiently diminished to halt the collapse. If this actually accounted for the enlargement of the damage zone, the restraint of the twisting deformations may have prevented the failure of column D-8.

Although a considerable amount of debris fell from the upper floors onto the first-floor extension to the north, a two-story deep pile of debris accumulated on the 8th floor. By one estimate, although the debris distributed some of its weight by bridging action, the net effect would have been a 500-percent increase in dead load moment for the supporting beam. Based on the computed results, and in the absence of wind, it appears that the connections would have been able to support more than 500 percent of the estimated dead load moments before any hinging would occur. This may explain why multiple stories of debris came to rest at the 8th floor without incurring additional damage to the structure.

Because column D-8 failed below the 16th floor, the beam-to-column moment connection was the single most significant structural feature that helped limit the damage. The portion of the building above the collapsed floors was held in place by frame action of the perimeter. Static elastic analyses of the moment frame show very high stress levels; however, there was a negligible deformation directly above the damaged structure. Furthermore, connections that enable the beams to develop some membrane capacity improve a structure's ability to arrest collapse. The typical floor beam end connections with their A307 bolts were overloaded in direct tension. High-strength bolts would have provided significantly greater tensile ability and possibly held more beams in place through catenary action. Inelastic analyses demonstrate the role of the weaker connections in the response of the structure. Finally, stronger column splices may have made it more difficult for the damaged column to separate from the upper column. Heavier column splices could have allowed the damaged column to function as a hanger and limit the amount of collapsed area, or they could have tended to pull more of the frame down.

6.9 Recommendations

It is difficult to draw conclusions and more detailed study is required to understand how the collapse was halted. As better descriptions of the structural details become available, the observed patterns of damage may provide useful information in the calibration of numerical simulation tools. Some issues requiring further study are:

1. Whether the observed damage in the column flange, and not at the beam flange, of the moment frames top connection plates is due to high restraint in the welds.

2. Why the bottom flange welded connection has typically failed at the fillet weld to beam interface and not at the fillet weld to seat plate interface.

3. The impact response of various moment-connected details.

4. Whether composite construction would reduce local collapse zones. (There were no shear connectors to provide composite action between the floor beams and slab. Composite construction would have increased the capacity of the members and may have dissipated more of the impact energy; however, it may have also pulled a greater extent of the adjoining regions into the collapse zone.)

5. Whether perimeter rebar in the slabs could improve the structural response by providing catenary action and tensile force resistance in the slabs to reduce local collapse zones.

6. Whether the partial-strength connections permitted members to break away from the structure, thereby limiting the extent of damage. (If the moment connections had been designed for the capacity of the sections [as opposed to fully rigid partial strength based on design load and stiffness requirements], the building performance is likely to have been different.)

7. Whether the collapse zone would have been limited if the spandrels on the 16th, 17th, and 18th floors had not been so grossly distorted through twisting.

6.10 References

Guy Nordenson and Associates and Simpson, Gumpertz, and Heger, Inc. 2001. *Draft Structural Engineering Evaluation for the New York Department of Design and Construction (DDC).* Prepared for LZA Technology. October.

Rudderman, James, Office of. 1971–1975. "Bankers Trust Plaza." Structural Drawings. New York, NY.

Smilow, Jeffrey. 2001. Personal account. Cantor Seinuk Group, Inc. New York, NY.

Therese McAllister
David Biggs
Edward M. DePaola
Dan Eschenasy
Ramon Gilsanz

7 *Peripheral Buildings*

7.1 Introduction

In addition to the WTC buildings and Bankers Trust building, a number of other buildings suffered damage from the projectiles and debris resulting from the deliberate aircraft impacts into WTC 1 and WTC 2 on September 11, 2001, and the resulting collapse of WTC 1, WTC 2, and WTC 7. As discussed in Chapter 1, Section 1.4, on September 12, 2001, the first round of building inspections were contracted for the New York City Department of Buildings (DoB) and the New York City Department of Design and Construction (DDC). This chapter is based on the field observations made by the Building Performance Study (BPS) Team, the Structural Engineers Association of New York (SEAoNY) summary presented below, and *Life Safety Reports* prepared by LAZ Technology/Thornton-Tomasetti (LZA 2001).

The building assessments were compared and coordinated with a parallel inspection performed by DoB and are summarized in Figure 7-1 and Table 7.1 according to the following color coding:

■	Green	Inspected	No significant damage found.
□	Yellow	Moderate Damage	Broken glass, façade damage, roof debris.
■	Blue	Major Damage	Damage to structural members requiring shoring or significant danger to occupants from glass, debris, etc.
■	Red	Partial Collapse	Building is standing, but a significant portion is collapsed. All of these buildings were inspected and found to have no remaining certifiable structural capacity.
■	Black	Full Collapse	Building is not standing.

It is important to note the distinction between evaluations for occupancy, access, and life safety and those for structural safety. Extensive damage to glazing and façades may pose a significant threat to the public, but there may be no structural damage. "Major Damage" has both of these types of damage. Because this report is concerned with building performance, primarily structural and fire performance, major damage categories that do not include structural or fire damage are not specifically addressed.

CHAPTER 7: *Peripheral Buildings*

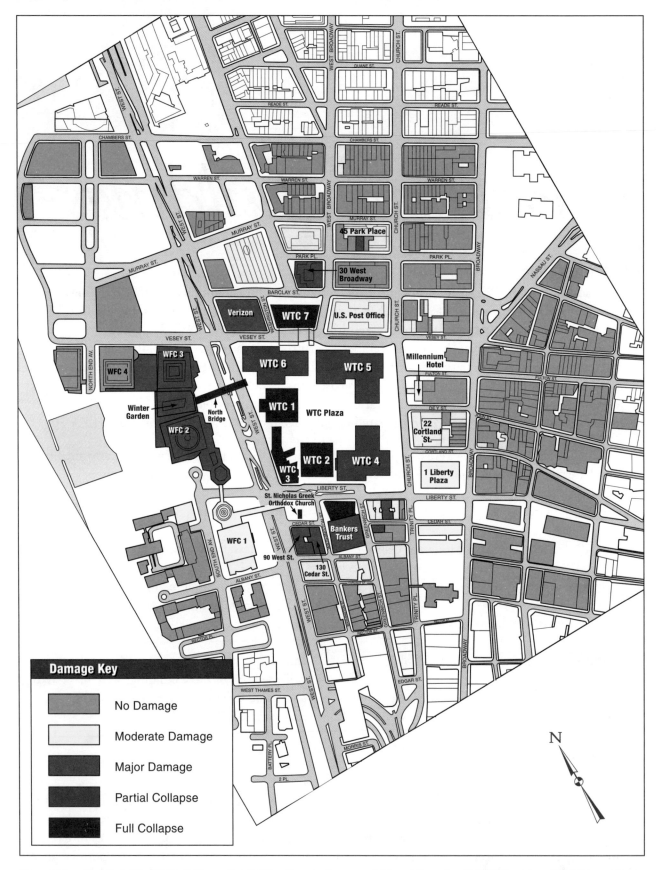

Figure 7-1   New York City DDC/DoB Cooperative Building Damage Assessment Map of November 7, 2001 (based on SEAoNY inspections of September and October 2001).

CHAPTER 7: *Peripheral Buildings*

Table 7.1 DoB/SEAoNY Cooperative Building Damage Assessment – November 7, 2001[1]

No.	Block	Lot	Address	Name	Building Color Code[2]	Building Rating
1	16	100	395 South End Ave.	Gateway	Yellow	Moderate Damage
2	16	120	120 West St.	1 WFC Tower A	Yellow	Moderate Damage
3	16	120	120 West St.	South Bridge	Yellow	Moderate Damage
4	16	120	120 West St.	1-2 WFC Link Bridge	Yellow	Moderate Damage
5	16	125	125 West St.	2 WFC Tower B	Blue	Major Damage
6	16	140	200 Vesey St.	3 WFC Tower C - Annex	Blue	Major Damage
7	16	140	201 Vesey St.	Winter Garden Building	Blue	Major Damage
8	48	1	2 Wall St.		Yellow	Moderate Damage
9	49	2	111 Broadway		Yellow	Moderate Damage
10	51	14	125 Greenwich St.		Yellow	Moderate Damage
11	51	15	90 Trinity Pl.		Yellow	Moderate Damage
12	52	10	120 Cedar St.		Blue	Major Damage
13	52	15	110 Trinity Pl.		Yellow	Moderate Damage
14	52	21	120 Liberty St.		Yellow	Moderate Damage
15	52	22	124 Liberty St.	Fire Station	Yellow	Moderate Damage
16	52	30	106 Liberty St.		Yellow	Moderate Damage
17	52	7501	110 Liberty St.		Yellow	Moderate Damage
18	52	7502	114 Liberty St.	Engineering Building	Blue	Major Damage
19	53	23	5 Carlisle		Yellow	Moderate Damage
20	53	28	1 Carlisle		Yellow	Moderate Damage
21	53	33	110 Greenwich St.		Yellow	Moderate Damage
22	54	1	130 Liberty St.	Bankers Trust	Blue	Major Damage
23	56	1	130 Cedar St.		Blue	Major Damage
24	56	20	155 Cedar St.	Greek Orthodox Church	Black	Collapse
25	56	4	90 West St.		Blue	Major Damage
26	58		WTC 1	North Tower	Black	Collapse
27	58		WTC 2	South Tower	Black	Collapse
28	58		WTC 3	Marriott International Hotel	Black	Collapse
29	58	1	WTC 4	South East Plaza	Red	Partial Collapse
30	58	1	WTC 5	North East Plaza	Red	Partial Collapse
31	58	1	WTC 6	Custom House	Red	Partial Collapse
32	84		WTC 7		Black	Collapse
33	62	1	1 Liberty Plaza		Yellow	Moderate Damage
34	63	1	10 Cortland St.		Yellow	Moderate Damage
35	63	3	22 Cortland St.		Yellow	Moderate Damage
36	63	6	27 Church St.	Century 21	Yellow	Moderate Damage
37	63	13	189 Broadway		Yellow	Moderate Damage
38	63	15	187 Broadway		Yellow	Moderate Damage
39	65	10	9 Maiden Ln.	Jeweler's Building	Yellow	Moderate Damage
40	65	16	174 Broadway		Yellow	Moderate Damage
41	80	4	47 Church St.	Millennium Hotel	Yellow	Moderate Damage
42	84	1	140 West St.	Verizon	Blue	Major Damage
43	86	1	90 Church St.	Post Office	Yellow	Moderate Damage
44	88	2	12 Vesey St.		Yellow	Moderate Damage
45	88	8	26 Vesey St.		Yellow	Moderate Damage
46	88	10	28 Vesey St.		Yellow	Moderate Damage
47	88	13	14 Barclay St.		Yellow	Moderate Damage
48	125	20	100 Church St.		Yellow	Moderate Damage
49	126	2	110 Church St.		Yellow	Moderate Damage
50	126	9	45 Park Pl.		Blue	Major Damage
51	126	27	120 Church St.		Yellow	Moderate Damage
52	127	1	30 West Broadway		Blue	Major Damage
53	127	18	75 Park Pl.		Yellow	Moderate Damage
54	128	2	224 Greenwich St.		Yellow	Moderate Damage
55	136	15	60 Warren St.		Yellow	Moderate Damage
56	136	16	128 Chambers St.		Yellow	Moderate Damage

[1] Adapted from SEAoNY inspections of September and October 2001 – Building Ratings and Actions table.

[2] Based on DDC/SEAoNY inspections of September and October 2001 and DoB inspections of October 22, 2001.

The following buildings suffered the most severe collateral damage from the collapse of the WTC towers:

Major Damage (shoring and large debris removal required):

WFC 3, American Express

Verizon

30 West Broadway

45 Park Place

Bankers Trust

90 West Street

130 Cedar Street

Partial Collapse

WTC 4

WTC 5

WTC 6

Winter Garden building (later revised to Major Damage)

Full Collapse

WTC 3

WTC 7

North Bridge from Winter Garden to WTC 1

St. Nicholas Greek Orthodox Church

Damage to the WTC and Bankers Trust buildings has been covered in previous chapters, and no inspection was made of St. Nicholas Greek Orthodox Church, because it was completely destroyed by falling debris. The damage to and performance of the remaining buildings is briefly presented. Immediately after the collapse of the towers, One Liberty Plaza was reported to be near collapse, but was later found to have no structural damage. The events leading up to this misunderstanding are briefly discussed.

7.2 World Financial Center

The World Financial Center (WFC) complex is located immediately west of the WTC Plaza, and includes four office towers, pedestrian walkways, and the Winter Garden (Figure 7-1). The buildings are of contemporary construction dating from 1985 to the present, have large floor plans, and are steel-framed structures with granite-clad curtain wall facades.

These buildings sustained varying degrees of façade and structural damage from the debris, with the eastern elevations experiencing the heaviest damages. The north and south elevations sustained lesser debris damage. WFC 1, 2, and 3 suffered glazing and façade damage, but WFC 4 was undamaged. Debris and dust penetrated nearly the full floor areas of WFC 2 and WFC 3 at several levels.

7.2.1 The Winter Garden

The Winter Garden lies between WFC 2 and WFC 3. It is a large greenhouse structure with a glass and steel telescopic barrel vault roof and is one of the largest public spaces in New York. The structure covers an area of approximately 200 feet by 270 feet and includes a public space of 120 feet by 270 feet. The largest vault has a clear height of about 130 feet and a span of 110 feet. The west elevation is made entirely of glass panels. The east end of the building has five composite steel floors that support a glass dome that

CHAPTER 7: Peripheral Buildings

covers a ceremonial stair. The structure has expansion joints where it meets WFC 2 and WFC 3. The spatial stability of the frame is insured by trussed arch framing. The east end was linked to the WTC complex by the North Bridge, which had a 200-foot clear span and was 40 feet wide. The west end has an entrance door.

Columns from WTC 1 hit the east end of the structure, particularly the area directly adjacent to the North Bridge. The Winter Garden experienced severe collapse of the eastern end framing. Several other semicircular trusses and parts of the dome were also badly damaged. The western two bays of the roof structure remained intact, but were covered with debris. Inspectors estimated that 60 percent of the roofing glass panels of the structure had collapsed.

Additional structural collapse occurred on parts of the 2nd and 3rd floor framing adjacent to WFC 2 and WFC 3, the North Bridge connection extension, the ceremonial stair above the circular landing, and the 4th and 5th floors at the eastern end. Localized structural collapse occurred in various other areas of the barrel roof.

As the eastern roof trusses were sheared in several places, support was provided only by the transverse plate girders that remained in place. These conditions, coupled with the shearing of trussed arch framing, led the first round of inspectors to conclude that the structure was potentially unstable, and a rating of Partial Collapse was assigned. After installation of shoring, a new evaluation of the building led engineers to determine that the building was repairable, and the rating was revised to Major Damage.

7.2.2 WFC 3, American Express Building

The 50-story WFC 3 building has a plan area of approximately 200 feet by 250 feet. Exterior column trees from WTC 1 were found hanging from the southeast corner of WFC 3 (Figure 7-2) and on the setback roof and against the east face of the Winter Garden (Figures 7-3 and 7-4). The impact of exterior column trees caused structural damage in both structures. Building faces not directly oriented toward the WTC site suffered minimal damage, even at the close proximity of several hundred yards.

Figure 7-2 Southeast corner of WFC 3.

FEDERAL EMERGENCY MANAGEMENT AGENCY

Figure 7-3
View of Winter Garden damage from West Street, with WTC 1 debris in front of WFC 2.

Figure 7-4
View of Winter Garden damage from West Street, with WTC 1 debris leaning against WFC 3.

The glazing and façade damage in the building was similar to that found in WFC 1 and WFC 2, which also had extensive cracking and breakage of glazing and granite panels. Debris from WTC 1 caused a collapse of the top 8 stories of the 10-story octagonal extension located at the southeast side of the building. The main WFC 3 building suffered damage from floors 17 to 26. A three-story section of exterior column trees from WTC 1 hung from the base of the collapsed area at floor 20, as shown in Figure 7-2, with approximately 25 feet of the column hanging outside the building. At floors 17 through 26, the corner column had been removed by the impact of debris, and the floors cantilevered from adjacent columns to the north and west. Smaller column debris penetrated floor 17. The damage did not extend past the corner bay, which had to be shored and was later demolished.

Interior damage is shown in Figure 7-5. Inspection of the interior determined that steel framing members that sustained direct impact from large debris had significant portions of the cementitious fireproofing material knocked off. The fireproofing was intact on adjacent steel members that had not been directly hit.

The localized nature of the damage, given the size of projectiles that impacted the building, is notable. Observations noted small welds between column end bearing plates at exterior and interior columns, indicating the columns near the damage zone were designed for gravity loads, and tension loads from wind were not a critical design parameter. This type of connection between columns may have allowed a column member to be knocked out of place without causing substantial displacement or damage to connecting framing.

7.3 Verizon Building

The Verizon building is located on the block bounded by Barclay Street on the north, Washington Street on the east, Vesey Street on the south, and West Street on the west. It is north of WTC 1 and WTC 2, and immediately west of WTC 7, which all collapsed.

Figure 7-5   Interior damage at floor 20 of WFC 3.

CHAPTER 7: Peripheral Buildings

The 30-story Verizon building was built in the 1930s and has a steel frame with infill exterior walls of unreinforced masonry, and five basement levels. The steel frame is encased in cinder-concrete and draped-wire-mesh, with cinder-concrete slab floor construction. Beams are rolled sections (mostly 12 inches deep) with cover plates at floors with high live loads. Girders are either rolled sections or built up from plates and angles. Columns are also built-up sections. Partially restrained moment frames at the building perimeter provide lateral resistance. The masonry walls are about 12 inches thick (on average), and the columns are encased in brick. The façade and 1st floor lobby are registered as historic landmarks. At the time of the adjacent building collapses, the Verizon building was in the midst of an extensive façade restoration program.

The proximity of the building to WTC 2 resulted in considerable damage to the south and east faces of the building. Damage included collapsed floor slabs and deformed beams and columns, including some local buckling. Window damage was moderate, and it is notable that the windows contained wire mesh. The west (West Street) and north (Barclay Street) sides of the building were not damaged.

The east (Washington Street) side of the building was damaged from about the 9th floor down, primarily due to the impact of debris sliding out from the base of WTC 7 (Figures 7-6, 7-7, and 7-8). Some damage may have also been caused by WTC 1 debris. In addition to fairly extensive façade damage (bricks and windows), there was damage to two bays of slab and framing at the 1st, 4th, and 7th floors and to one bay of slab and framing (including spandrel beam) at the 1st floor mezzanine and at the 5th floor. Two exterior columns suffered major damage between the 1st and 2nd floors (Figure 7-9), one exterior column suffered minor damage between the 3rd and 5th floors, and two exterior columns suffered major damage between the 6th and 8th floors. In addition, one interior column suffered minor damage below the 7th floor.

Figure 7-6
Verizon building – damage to east elevation (Washington Street).

CHAPTER 7: *Peripheral Buildings*

Figure 7-7   Verizon building – damage to east elevation (Washington Street) due to WTC 7 framing leaning against the building.

Figure 7-8   Verizon building – damage to east elevation (Washington Street).

CHAPTER 7: *Peripheral Buildings*

Figure 7-9   Verizon building – column damage on east elevation (Washington Street).

The south (Vesey Street) side of the building was damaged from approximately the 13th floor down, primarily due to the impact of projectile debris from the collapse of WTC 1 (Figure 7-10). In addition to fairly extensive façade damage (bricks and windows), two bays of slab and framing were damaged at the sidewalk arcade at the 1st floor, and one bay of slab and framing (including spandrel beams) was damaged at the 6th, 7th, 9th, 11th, 12th, and 13th floors (Figures 7–11 and 7-12). In addition, one interior column suffered minor damage below the 1st floor.

None of the damage to the floor framing threatened the structural integrity of the building. Although the damaged columns were deflected out-of-plane, it was determined that the columns were stable and not in danger of imminent collapse.

In general, the Verizon building performed well, especially given its close proximity to WTC 7. On the south (Vesey Street) side of the building, damage was extremely localized near the point of impact of projectile debris. In some cases, only a short section of spandrel beam and small area of floor slab were damaged, leaving the remainder of the structural bay intact (Figure 7-12).

On the east (Washington Street) side of the building, most of the damage appeared to be due to the lateral pressure of the spreading debris at the base of WTC 7 (Figure 7-7). Two of the columns between the 1st and 2nd floors were deflected into the building by as much as 2 feet (with most of the rotation occurring at the column splice just above the 1st floor); at one of the columns, very little contact remained at the column splice (Figure 7-9). Even so, the columns did not buckle, and structural bays above did not collapse or deflect significantly. Similarly, the structural bays supported by the column between the 6th and 8th floors that was completely destroyed by the impact of projectile debris were essentially undamaged.

CHAPTER 7: *Peripheral Buildings*

*Figure 7-10
Verizon building – damage to south elevation (Vesey Street).*

Figure 7-11 Verizon building – localized damage to south elevation (Vesey Street).

CHAPTER 7: *Peripheral Buildings*

Figure 7-12 Verizon building – detail of damage to south elevation (Vesey Street).

Several factors may have contributed to the performance of the Verizon building. The thick masonry walls, brick-encased columns, and cinder-concrete-encased beams and girders probably absorbed much of the energy of the impacts while also providing additional stiffness and strength to the building frame. The lower floors (up to the 10th floor) were designed for either 150 pounds per square foot (psf) or 275 psf, depending on the intended occupancy. Consequently, the member sizes and end connections are unusually stocky. Although designed for higher-than-normal live loads, at the time of the adjacent building collapses, actual live loads were relatively low.

Most floors are framed with 12-inch-deep beams with cover plates, presumably to maintain uniform floor clearances. These sections had full lateral bracing and were probably able to develop close to their full plastic capacities without buckling. Almost every beam-column connection was nominally moment-resistant, making the structural system highly redundant. All of these characteristics combined to both absorb the energy of the debris impacts and provide alternate load paths around the damaged areas.

The performance of this building led to several observations. The original design was for a substantially heavier live load and use as a telephone switching facility. Even so, the exterior columns on the east side were substantially damaged at the lower floors by the collapse of WTC 7. The nominally 12-inch thick brick masonry perimeter walls absorbed a significant portion of the impact energy, resulting in less damage to the structural steel framing. Impact damage was localized and did not propagate beyond immediate points of impact (sometimes not even full bays were damaged). It was noted that the windows performed better than those in other peripheral buildings, likely due to the wire mesh.

7.4 30 West Broadway

The office structure at 30 West Broadway is most recently known as Fiterman Hall of the Borough of Manhattan Community College campus of the City University of New York. It is located just north of WTC 7. The 17-story building was constructed in the 1950s and has a concrete-encased structural steel frame with cinder-concrete floor slabs with draped steel mesh. The structure had riveted, bolted, and welded connections, and roof setbacks at the 6th and 15th floor levels. The curtain wall consists of horizontal bands of windows over glazed brick. There are continuous lintels at every floor. The building was in the final stages of rehabilitation work at the time of the terrorist attacks.

The southern half of the west façade and most of the south façade were severely damaged or destroyed. The south face of the building suffered structural damage in the exterior bay from impact by large debris from WTC 7 (Figure 7-13). There was no damage to the east and north faces of the building, and no fire in the building, even though there was a substantial fire in WTC 7.

Figure 7-13
30 West Broadway – south façade, 6th floor to roof, looking northeast.

Damage was concentrated along the south face at and below the setback at the 15th floor. Portions of the south façade from the 15th floor collapsed. A vertical section of the perimeter wall extending five floors down from the setback at the center of the south façade was raked away. Local collapse also occurred at the southwest corner. The majority of the glass panes were knocked out on the south façade, in a triangular pattern that extended to the full width of the base. The south side of the building was unstable and required bracing. Floors 9 through 14 had two collapsed bays, and floors 3 through 6 had up to three collapsed bays. No structural damage was observed one bay away from the impact damage.

Floors 9 through 14 had at least two collapsed exterior bays and floors 3 through 6 had at least three collapsed exterior bays. There was relatively little damage at the 7th floor. A considerable amount of debris was on the 8th floor. The steel beams supporting two bays of this floor yielded, but are still in place.

The building was impacted by debris from the collapse of WTC 7. Although structural damage from debris impact was contained to the exterior bays on the south side of the building and between roof setback levels, it was more extensive than that observed on the east side of the Verizon building.

7.5 130 Cedar Street

Constructed in the 1930s, the building at 130 Cedar Street is a 12–story reinforced concrete frame structure with setbacks at the 10th, 11th, and 12th floors (Figure 7-14). The building is directly east of 90 West Street and is bordered by Cedar Street to the north, Washington Street to the east, and Albany Street to the south.

The floor framing consists of reinforced concrete flat slabs supported on square columns with capitals and dropped slabs. Columns are spaced at approximately 16 feet on center in the east-west direction and approximately 21 feet on center in the north-south direction. Perimeter concrete spandrel

Figure 7-14 130 Cedar Street and 90 West Street.

beams beneath the windows and interior infill walls of brick, terra cotta, or concrete masonry provide additional lateral stiffness.

Some façade damage was noted (primarily to the parapets), but most of the damage occurred at the roof level where the slab of the northeast corner collapsed under debris with the column capitals punching through the slab. A column section from WTC 2 penetrated the 10th floor roof slab. The southern portion of the building was not damaged. Structural damage from projectile impact and fire occurred primarily above the 9th floor. Fire damage was evident on the 11th and 12th floors in the northwest corner. Several concrete columns were cracked, possibly from the impact. Several bays at the northeast corner were severely damaged from debris impact. Concrete samples from two fire locations indicated that the concrete structure may have experienced fire temperatures of between 315 °C (600 °F) and 590 °C (1,100 °F). Spalling of capitals was observed in the fire areas.

The masonry infill walls were cracked throughout the building. It is not clear whether the condition pre-existed, or if it was due to the fire, floor settlement, or frame movement.

7.6 90 West Street

This building is located south of the WTC site, and adjacent to the 130 Cedar Street building located on the west side, as shown in Figure 7-14. The 24-story building has a steel-frame structure with a terra cotta flat-arch floor system and infill walls of unreinforced masonry. It was designed by architect Cass Gilbert and structural engineer Gunvald Aus in 1907. The floor plan has a skewed "C" configuration, with overall dimensions of approximately 124 feet by 180 feet. At the higher floors, the typical exterior wall assembly is terra cotta tiles on a brick wall. This building is a designated New York City landmark. In 1907, the building towered over the waterfront and warehouses in the area. Its top floor had a restaurant that was billed as the "world's highest."

The riveted steel framing consists of rolled and built-up sections for the columns and beams. Columns are spaced approximately 18 feet apart. The primary framing runs north-south, with secondary members in the east-west direction. Lateral load resistance is provided primarily by partial-strength, partially restrained moment connections of frame members and the infill masonry walls. The floor slabs of terra cotta flat arches appeared to be topped with low-strength cinder-concrete. Terra cotta and masonry enclosures provided fireproofing for all original architectural areas and structural elements. The building construction is shown in Figure 7-15.

The New York City Building Code required the floors to be tested for 4 hours while exposed to a fire maintained at 927 °C (1,700 °F) and a load of 150 psf. Following the fire test, the fire-exposed underside was exposed to a fire hose stream with a nozzle pressure of 60 pounds per square inch (psi) for 10 minutes. The floor was then loaded and unloaded with a uniform load of 600 psf in the middle bay. The test was considered successful if the deflection of the beams supporting the assembly was less than 2.5 inches over a 14-foot length, after cooling.

The building was undergoing façade rehabilitation and was fully covered with scaffolding. Many of the interior columns still had the original terra cotta covers, and some were covered with plaster, but others were covered with sheet rock and intumescent paint, and, at one location, there was a metal deck with spray-on fireproofing. In some locations, spray-on cementitious fireproofing was used for later tenant work. Some scaffolding planks caught fire and may have contributed to the spread of the fire between floors.

Terra cotta and hollow-clay tile arches were common in fireproof office construction. Most of them were patented systems with 6- to 15-inch depths and spans from 54 to 90 inches. The arches were supported on the bottom flanges of steel beams. The bottom flanges of the supporting steel beams were generally

CHAPTER 7: Peripheral Buildings

Figure 7-15 Interior of 90 West Street showing typical construction features.

protected by clay tile or terra cotta fireproofing. To provide a smooth finish, the arches were usually topped with a cementitious material that also protected the haunches of the steel beams. The arches had tie rods to resist the thrust of the arch. An 8-inch flat arch with hollow tile and a span of 6 feet could carry a safe load of 170 psf (Kidder 1936). At 90 West Street, the tile floor arches usually span 6 feet. At lower floors, the tiles have a 12-inch thickness and cover the bottom flanges of the beams.

The roof was damaged by debris falling from WTC 2, and approximately half of the north face of the building experienced projectile impact and fire damage. WTC 2 projectiles severed spandrel beams at floors 8 to 11 in the 2nd bay from the west end, and in a middle bay at the 6th floor. Terra cotta slabs were damaged mostly in the exterior bay at these locations. In general, the projectiles damaged only the masonry and broke many terra cotta features. The damage to the interior structural terra cotta floor slabs was primarily due to the brittle fracture of the terra cotta slabs upon impact by large debris. Most of the damage was restricted to the two northernmost bays, with the exception of fire damage on the 1st through 5th, 7th through 10th, 14th, 21st, and 23rd floors. The fire did not spread to the south side of the building, except for the first 4 floors. Columns were buckled 1-2 inches on the 8th and 23rd floors, approximately a foot below the ceiling, as shown in Figures 7-16 and 7-17. A tube column supporting a north exit stair from the roof and a built-up column supporting the roof were the only other heat-induced buckling damage observed during initial inspections.

This type of construction, with terra cotta tiles providing fire protection, was common in early 20th century construction. The style of construction resulted in a highly compartmentalized building, which may have helped slow the spread of fire. The Fire Department of New York was able to control the fires in this building. The fire damage observed in the building, with minimal structural damage from a normal

Figure 7-16
Buckling damage at top of column on floor 8 of 90 West Street. Note the loss of fire protection at the top of the column.

fire load, is considered typical for this type of construction and fire protection; however, it has been suggested that the scaffolding that was in place for renovations contributed to the spread of fire between floors that may not have occurred otherwise. However, the only structural damage observed was buckling damage near the tops of two columns.

7.7 45 Park Place

This building is located three blocks north of the WTC site (Figure 7-1), and was initially rated as No Damage when inspected from the exterior. However, subsequent interior inspection revealed that three floor beams were missing from the top story of the building as a result of the landing gear that penetrated the roof following the airplane impact on WTC 2, shown in Figure 1-4 (in Chapter 1). The rating was subsequently changed to Major Damage. No other significant damage was found.

7.8 One Liberty Plaza

One Liberty Plaza (One Liberty) is a 54-story, 730-foot-high building, comprising a footprint of 238 feet in the north-south direction by 163 feet in the east-west direction. The building area is approximately 2,000,000 square feet (Figures 7-18 and 7-19). It was designed by Skidmore, Owings, and Merrill in 1970 and served as corporate headquarters for U. S. Steel Corporation.

CHAPTER 7: *Peripheral Buildings*

Figure 7-17 Buckling damage at top of column on floor 23 of 90 West Street.

During the afternoon of September 11, following collapse of WTC 1 and WTC 2, rumors were spread that One Liberty was in imminent danger of collapse. This was due to a report by an untrained observer that the building face appeared to be moving or leaning.

The majority of damage incurred at One Liberty consisted of broken window glass and frames. Most of that broken glass was in the lower six floors of the west-facing elevation, with less breakage on the floors above. There was, however, some broken glass on the north- and south-facing elevations as well. At those elevations, most of the broken glass was located at floors 1 through 6. There were approximately 550 broken lites of glass, and approximately 200 frames were damaged beyond repair.

On September 12, there was a persistent rumor that One Liberty was still in danger of collapse. The building was inspected by structural engineers conducting building surveys and safety evaluations of buildings around the WTC site. The building vertical alignment was measured with a transit to determine whether any lateral drift had occurred along the height of the building. Three locations on the west face were evaluated, and no apparent movement was observed in the building. The One Liberty Plaza building was determined to be safe, except for dangers related to broken glass.

Statements were released about the safety of the building, but the rumors persisted on September 13. To stop the rumors and convince the public of the building's safety, DDC surveyors continuously monitored the building and engineers inspected each floor. When a piece of glass fell off One Liberty during the afternoon, it was rumored to be a partial collapse. The findings of the engineers and surveyors concerning the structural safety of One Liberty were reported on the nightly news, and the rumors stopped.

Figure 7-18 One Liberty Plaza – south elevation, lower floors.

7.9 Observations and Findings

Steel-frame construction from the 1900s through the 1980s, though different in many details, performed well under significant impact loads by limiting impact damage and progressive collapse to local areas.

Heavy unreinforced masonry façades were observed to absorb significant amounts of impact energy in the Verizon and 90 West Street buildings. Heavy masonry façades like those in the Verizon, 90 West Street, or even 130 Cedar Street buildings may also provide an alternative load path for a damaged structure.

Older, early-century fireproofing methods of concrete-, brick-, and terra cotta tile-encased steel frames performed well, even after 90+ years, and protected the 90 West Street building from extensive structural damage.

7.10 Recommendations

The known data and conditions of the perimeter structures after the impact damage should be utilized as a basis for calibration, comparison, and verification of existing software intended to predict such behavior, and for the development of new software for the prediction of the ability of structures to sustain localized and global overload conditions.

Figure 7-19
One Liberty Plaza – south elevation, upper floors.

7.11 References

Kidder-Parker. 1936. *Kidder-Parker Architects and Builders Handbook*. John Wiley & Sons, Inc. Section on Fire-Resistive Floor Construction.

LZA/Thornton-Tomasetti, Guy Nordenson and Associates, and Simpson, Gumpertz, and Heger, Inc. *Interim Life Safety Report, 90 West Street, New York, NY, November 15, 2001*. Prepared for New York City Department of Design and Construction.

LZA/Thornton-Tomasetti and Gilsanz Murray Steficek LLP. *Interim Life Safety Report, 30 West Broadway, New York, NY, November 21, 2001*. Prepared for New York City Department of Design and Construction.

LZA Technology/Thornton-Tomasetti, Structural Engineers Association of New York, and Guy Nordenson, et al., editors. Est. April 2002. *WTC Emergency Damage Assessment of Buildings, Structural Engineers Association of New York Inspections of September and October 2001*. Draft prepared for the New York City Department of Design and Construction. New York, NY.

8 Observations, Findings, and Recommendations

The observations, findings, and recommendations of this building performance study are summarized in three sections:

- Summary of Report Observations, Findings, and Recommendations
- Chapter Observations, Findings, and Recommendations
- Building Performance Study (BPS) Recommendations for Future Studies

8.1 Summary of Report Observations, Findings, and Recommendations

In the study of the WTC towers and the surrounding buildings that were subsequently damaged by falling debris and fire, several issues were found to be critical to the observed building performance in one or more buildings.

These issues fall into several broad topics that should be considered for buildings that are being evaluated or designed for extreme events. It may be that some of these issues should be considered for all buildings; however, additional studies are required before general recommendations, if any, can be made for all buildings. The issues identified from this study of damaged buildings in or near the WTC site have been summarized into the following points:

a. Structural framing systems need redundancy and/or robustness, so that alternative paths or additional capacity are available for transmitting loads when building damage occurs.

b. Fireproofing needs to adhere under impact and fire conditions that deform steel members, so that the coatings remain on the steel and provide the intended protection.

c. Connection performance under impact loads and during fire loads needs to be analytically understood and quantified for improved design capabilities and performance as critical components in structural frames.

d. Fire protection ratings that include the use of sprinklers in buildings require a reliable and redundant water supply. If the water supply is interrupted, the assumed fire protection is greatly reduced.

e. Egress systems currently in use should be evaluated for redundancy and robustness in providing egress when building damage occurs, including the issues of transfer floors, stair spacing and locations, and stairwell enclosure impact resistance.

CHAPTER 8: *Observations, Findings, and Recommendations*

f. Fire protection ratings and safety factors for structural transfer systems should be evaluated for their adequacy relative to the role of transfer systems in building stability.

8.2 Chapter Observations, Findings, and Recommendations

The following sections present observations, findings, and recommendations specifically made in each chapter of this report, including the discussion of building codes and fire standards in Chapter 1 and the limited metallurgical examination of steel from the WTC towers and WTC 7 in Appendix C.

8.2.1 Chapter 1: Building Codes and Fire Standards

Observations and Findings

a. The decision to include aircraft impact as a design parameter for a building would clearly result in a major change in the design, livability, usability, and cost of buildings. In addition, reliably designing a building to survive the impact of the largest aircraft available now or in the future may not be possible. These types of loads and analyses are not suitable for inclusion in minimum loads required for design of all buildings. Just as the possibility of a Boeing 707 impact was a consideration in the original design of WTC 1 and WTC 2, there may be situations where it is desirable to evaluate building survival for impact of an airplane of a specific size traveling at a specific speed. Although there is limited public information available on this topic, interested building owners and design professionals would require further guidance for application to buildings.

b. The ASTM E119 Standard Fire Test was developed as a comparative test, not a predictive one. In effect, the Standard Fire Test is used to evaluate the relative performance (fire endurance) of different construction assemblies under controlled laboratory conditions, not to predict performance in real, uncontrolled fires.

8.2.2 Chapter 2: WTC 1 and WTC 2

8.2.2.1 Observations and Findings

a. The structural damage sustained by each of the two buildings as a result of the terrorist attacks was massive. The fact that the structures were able to sustain this level of damage and remain standing for an extended period of time is remarkable and is the reason that most building occupants were able to evacuate safely. Events of this type, resulting in such substantial damage, are generally not considered in building design, and the ability of these structures to successfully withstand such damage is noteworthy.

b. Preliminary analyses of the damaged structures, together with the fact the structures remained standing for an extended period of time, suggest that, absent other severe loading events such as a windstorm or earthquake, the buildings could have remained standing in their damaged states until subjected to some significant additional load. However, the structures were subjected to a second, simultaneous severe loading event in the form of the fires caused by the aircraft impacts.

c. The large quantity of jet fuel carried by each aircraft ignited upon impact into each building. A significant portion of this fuel was consumed immediately in the ensuing fireballs. The remaining fuel is believed either to have flowed down through the buildings or to have burned off within a few minutes of the aircraft impact. The heat produced by this burning jet fuel does not by itself appear to have been sufficient to initiate the structural collapses. However, as the burning jet fuel spread across several floors of the buildings, it ignited much of the buildings' contents, causing simultaneous fires across several floors of both buildings. The heat output from these fires is estimated to have been

CHAPTER 8: *Observations, Findings, and Recommendations*

comparable to the power produced by a large commercial power generating station. Over a period of many minutes, this heat induced additional stresses into the damaged structural frames while simultaneously softening and weakening these frames. This additional loading and the resulting damage were sufficient to induce the collapse of both structures.

d. Because the aircraft impacts into the two buildings are not believed to have been sufficient to cause collapse without the ensuing fires, the obvious question is whether the fires alone, without the damage from the aircraft impact, would have been sufficient to cause such a collapse. The capabilities of the fire protection systems make it extremely unlikely that such fires would develop without some unusual triggering event like the aircraft impact. For all other cases, the fire protection for the tower buildings provided in-depth protection. The first line of defense was the automatic sprinkler protection. The sprinkler system was intended to respond quickly and automatically to extinguish or confine a fire. The second line of defense consisted of the manual (FDNY/Port Authority Fire Brigade) firefighting capabilities, which were supported by the building standpipe system, emergency fire department use elevators, smoke control system, and other features. Manual suppression by FDNY was the principal fire protection mechanism that controlled a large fire that occurred in the buildings in 1975. Finally, the last line of defense was the structural fire resistance. The fire resistance capabilities would not be called upon unless both the automatic and manual suppression systems just described failed. In the incident of September 11, not only did the aircraft impacts disable the first two lines of defense, they also are believed to have dislodged fireproofing and imposed major additional stresses on the structural system.

e. Had some other event disabled both the automatic and manual suppression capabilities and a fire of major proportions occurred while the structural framing system and its fireproofing remained intact, the third line of defense, structural fireproofing, would have become critical. The thickness and quality of the fireproofing materials would have been key factors in the rate and extent of temperature rise in the floor trusses and other structural members. In the preparation of this report, there has not been sufficient analysis to predict the temperature and resulting change in strength of the individual structural members in order to approximate the overall response of the structure. Given the redundancy in the framing system and the capability of that system to redistribute load from a weakened member to other parts of the structural system, it is impossible, without extensive modeling and other analysis, to make a credible prediction of how the buildings would have responded to an extremely severe fire in a situation where there was no prior structural damage. Such simulations were not performed within the scope of this study, but should be performed in the future.

f. Buildings are designed to withstand loading events that are deemed credible hazards and to protect the public safety in the event such credible hazards are experienced. Buildings are not designed to withstand any event that could ever conceivably occur, and any building can collapse if subjected to a sufficiently extreme loading event. Communities adopt building codes to help building designers and regulators determine those loading events that should be considered as credible hazards in the design process. These building codes are developed by the design and regulatory communities themselves, through a voluntary committee consensus process. Prior to September 11, 2001, it was the consensus of these communities that aircraft impact was not a sufficiently credible hazard to warrant routine consideration in the design of buildings and, therefore, the building codes did not require that such events be considered in building design. Nevertheless, the design of WTC 1 and WTC 2 did include at least some consideration of the probable response of the buildings to an aircraft impact, albeit a somewhat smaller and slower moving aircraft than those actually involved in the September 11 events. Building codes do consider fire as a credible hazard and include extensive requirements to control the spread of fire throughout buildings, to delay the onset of fire-induced structural collapse, and to facilitate the safe egress of building occupants in a fire event. For fire-protected steel-frame buildings, like WTC 1 and WTC 2, these code requirements had been deemed effective and, in fact, prior to September 11, there was no record of the fire-induced-collapse of such structures, despite some very large uncontrolled fires.

g. The ability of the two towers to withstand aircraft impacts without immediate collapse was a direct function of their design and construction characteristics, as was the vulnerability of the two towers to collapse a result of the combined effects of the impacts and ensuing fires. Many buildings with other design and construction characteristics would have been more vulnerable to collapse in these events than the two towers, and few may have been less vulnerable. It was not the purpose of this study to assess the code-conformance of the building design and construction, or to judge the adequacy of these features. However, during the course of this study, the structural and fire protection features of the buildings were examined. The study did not reveal any specific structural features that would be regarded as substandard, and, in fact, many structural and fire protection features of the design and construction were found to be superior to the minimum code requirements.

h. Several building design features have been identified as key to the buildings' ability to remain standing as long as they did and to allow the evacuation of most building occupants. These included the following:

- robustness and redundancy of the steel framing system

- adequate egress stairways that were well marked and lighted

- conscientious implementation of emergency exiting training programs for building tenants

i. Similarly, several design features have been identified that may have played a role in allowing the buildings to collapse in the manner that they did and in the inability of victims at and above the impact floors to safely exit. These features should not be regarded either as design deficiencies or as features that should be prohibited in future building codes. Rather, these are features that should be subjected to more detailed evaluation, in order to understand their contribution to the performance of these buildings and how they may perform in other buildings. These include the following:

- the type of steel floor truss system present in these buildings and their structural robustness and redundancy when compared to other structural systems

- use of impact-resistant enclosures around egress paths

- resistance of passive fire protection to blasts and impacts in buildings designed to provide resistance to such hazards

- grouping emergency egress stairways in the central building core, as opposed to dispersing them throughout the structure

j. During the course of this study, the question of whether building codes should be changed in some way to make future buildings more resistant to such attacks was frequently explored. Depending on the size of the aircraft, it may not be technically feasible to develop design provisions that would enable all structures to be designed and constructed to resist the effects of impacts by rapidly moving aircraft, and the ensuing fires, without collapse. In addition, the cost of constructing such structures might be so large as to make this type of design intent practically infeasible.

Although the attacks on the World Trade Center are a reason to question design philosophies, the BPS Team believes there are insufficient data to determine whether there is a reasonable threat of attacks on specific buildings to recommend inclusion of such requirements in building codes. Some believe the likelihood of such attacks on any specific building is deemed sufficiently low to not be considered at all. However, individual building developers may wish to consider design provisions for improving redundancy and robustness for such unforeseen events, particularly for structures that, by nature of their design or occupancy, may be especially susceptible to such incidents. Although some conceptual changes to the building codes that could make buildings more resistant to fire or impact damage or more conducive to occupant egress were identified in the course of this study, the BPS Team felt that extensive technical, policy, and economic study of these concepts should be performed before any specific code change recommendations are developed. This report specifically recommends such additional studies. Future building code revisions may be considered after the technical details of the collapses and other building responses to damage are better understood.

8.2.2.2 Recommendations

The scope of this study was not intended to include in-depth analysis of many issues that should be explored before final conclusions are reached. Additional studies of the performance of WTC 1 and WTC 2 during the events of September 11, 2001, and of related building performance issues should be conducted. These include the following:

a. During the course of this study, it was not possible to determine the condition of the interior structure of the two towers, after aircraft impact and before collapse. Detailed modeling of the aircraft impacts into the buildings should be conducted in order to provide understanding of the probable damage state immediately following the impacts.

b. Preliminary studies of the growth and heat flux produced by the fires were conducted. Although these studies provided useful insight into the buildings' behavior, they were not of sufficient detail to permit an understanding of the probable distribution of temperatures in the buildings at various stages of the event and the resulting stress state of the structures as the fires progressed. Detailed modeling of the fires should be conducted and combined with structural modeling to develop specific failure modes likely to have occurred.

c. The floor framing system for the two towers was complex and substantially more redundant than typical bar joist floor systems. Detailed modeling of these floor systems and their connections should be conducted to understand the effects of localized overloads and failures to determine ultimate failure modes. Other types of common building framing should also be examined for these effects.

d. The fire-performance of steel trusses with spray-applied fire protection, and with end restraint conditions similar to those present in the two towers, is not well understood, but is likely critical to the building collapse. Studies of the fire-performance of this structural system should be conducted.

e. Observation of the debris generated by the collapse of the towers and of damaged adjacent structures suggests that spray-applied fireproofing may be vulnerable to mechanical damage from blasts and impacts. This vulnerability is not well understood. Tests of these materials should be conducted to understand how well they withstand such mechanical damage and to determine whether it is appropriate and feasible to improve their resistance to such damage.

f. In the past, tall buildings have occasionally been damaged, typically by earthquakes, and experienced collapse within the damaged zones. Those structures were able to arrest collapse before they progressed to a state of total collapse. The two WTC towers were able to arrest collapse from the impact damage, but not from the resulting fires when combined with the impact effects of the aircraft attacks. Studies should be conducted to determine, given the great size and weight of the two towers, whether there are feasible design and construction features that would permit such buildings to arrest or limit a collapse, once it began.

CHAPTER 8: *Observations, Findings, and Recommendations*

8.2.3 Chapter 3: WTC 3

8.2.3.1 Observations

WTC 3 was subjected to extraordinary loading from the impact and weight of debris from the two adjacent 110-story towers. It is noteworthy that the building resisted both horizontal and vertical progressive collapse when subjected to debris from WTC 2. The overloaded portions were able to break away from the rest of the structure without pulling it down, and the remaining structural system was able to remain stable and support the debris load. The structure was even capable of protecting occupants on lower floors after the collapse of WTC 1.

8.2.3.2 Recommendations

WTC 3 should be studied further to understand how it resisted progressive collapse.

8.2.4 Chapter 4: WTC 4, 5, and 6

WTC 4, 5, and 6 have similar design features, although their building configurations are somewhat different. Because WTC 5 was the only building accessible for observation, most of the following discussion focuses on this building. However, the observations, findings, and recommendations are assumed to be applicable to all three buildings.

8.2.4.1 Observations and Findings

a. All three buildings suffered extensive fire and impact damage and significant partial collapse. The condition of the stairways in WTC 5 indicates that, for the duration of this fire, the fire doors and the fire protective covering on the walls performed well. There was, however, damage to the fire side of the painted fire doors, and the damage-free condition on the inside or stairwell side of those same doors indicates the doors performed as specified for the fire condition that WTC 5 experienced. These stairway enclosures were unusual for buildings that have experienced fire because they were not impacted by water from firefighting operations. In addition, the stairway doors were not opened during the fire and remained latched and closed throughout the burnout of the floors. Therefore, general conclusions regarding the effectiveness of this type of stairway construction may not be warranted.

b. The steel generally behaved as expected given the fire conditions in WTC 5. Many beams developed catenary action as illustrated in Figure 4-14. Some columns buckled, as shown in Figure 4-17. The one exception is the limited internal structural collapse in WTC 5. The fire-induced failure that led to this collapse was unexpected. As in the rest of the building, the steel beams were expected to deflect significantly, yet carry the load. This was not the case where the beam connections failed. The failure most likely occurred during the heating of the structure because the columns remained straight and freestanding after the collapse.

c. The structural redundancy provided by the exterior wall pipe columns helped to support the cantilevered floors. This was important because it kept the cantilevers from buckling near the columns as might be expected.

d. The limited structural collapse in WTC 5 due to fire impact as described in Section 4.3.2 appeared to be caused by a combination of excessive shear loads and tensile forces acting on the simple shear connections of the infill beams. The existence of high shear loads was evident in many of the column tree beam stub cantilevers that formed diagonal tension field failure mechanisms in the cantilever webs, as seen in Figure 4-19.

e. The end bearing resistance of the beam web was less than the double shear strength of the high-strength bolts. An increased edge distance might have prevented this collapse by increasing the

CHAPTER 8: *Observations, Findings, and Recommendations*

connections' tensile strength. The failure most likely began on the 8th floor and progressed downward, because the 9th floor did not collapse. The 4th floor and those below remained intact.

f. The 7th floor framing was shop-coated. In some locations, the paint appeared to be in good condition and not discolored by the fire. Paint usually blisters and chars when heated to temperatures of about 100 °C (212 °F). This indicates that the fire protection material remained on the steel during the early phase of the fire and may have fallen off relatively later in the fire as the beams twisted, deflected, and buckled. Additional measures for proper adhesion may be required when applying spray-on fire protection to painted steel.

g. On the lower floors, the steel beams appeared to have heat damage from direct fire impact and there was little or no evidence of shop painting, indicating that fireproofing material was either missing before the fire or delaminated early in the fire exposure.

h. In general, the buildings responded as expected to the impact loadings. Collapse was often localized, although half of WTC 4 and most of the central part of WTC 6 suffered collapse on all floors. The damage was consistent with the observed impact load.

i. Reinforced web openings in steel beams performed well, as no damage or local buckling was observed at these locations.

j. The automatic sprinkler system did not control the fires. Some sprinkler heads fused, but there was no evidence of significant water damage, due to a lack of water. This is consistent with the lack of water damage in the bookstore on the lower level and the complete burnout of the upper floors.

8.2.4.2 Recommendations

The scope of this study and the limited time allotted prevented in-depth analysis of many issues that should be explored before final conclusions are reached. Additional studies of the performance of WTC 4, 5, and 6 during the events of September 11, 2001, and related building performance issues should be conducted. These include the following:

a. There is insufficient understanding of the performance of connections and their adequacy under real fire exposures as discussed in Appendix A. This is an area that needs further study. The samples discussed in Section 4.3.2 should be useful in such a study.

b. A determination of the combined structural and fire properties of the critical structural connections should be made to permit prediction of their behavior under overload conditions. This can be accomplished with a combination of thermal transfer modeling, structural finite element modeling (FEM), and full-scale physical testing.

8.2.5 Chapter 5: WTC 7
8.2.5.1 Observations and Findings

a. This office building was built over an electrical substation and a power plant, comparable in size to that operated by a small commercial utility. It also stored a significant amount of diesel oil and had a structural system with numerous horizontal transfers for gravity and lateral loads.

b. The loss of the east penthouse on the videotape suggests that the collapse event was initiated by the loss of structural integrity in one of the transfer systems. Loss of structural integrity was likely a result of weakening caused by fires on the 5th to 7th floors. The specifics of the fires in WTC 7 and how they caused the building to collapse remain unknown at this time. Although the total diesel fuel on the premises contained massive potential energy, the best hypothesis has only a low probability of occurrence. Further research, investigation, and analyses are needed to resolve this issue.

FEDERAL EMERGENCY MANAGEMENT AGENCY

CHAPTER 8: *Observations, Findings, and Recommendations*

c. The collapse of WTC 7 was different from that of WTC 1 and WTC 2. The towers showered debris in a wide radius as their external frames essentially "peeled" outward and fell from the top to the bottom. In contrast, the collapse of WTC 7 had a relatively small debris field because the façade came straight down, suggesting an internal collapse. Review of video footage indicates that the collapse began at the lower floors on the east side. Studies of WTC 7 indicate that the collapse began in the lower stories, either through failure of major load transfer members located above an electrical substation structure or in columns in the stories above the transfer structure. Loss of strength due to the transfer trusses could explain why the building imploded, with collapse initiating at an interior location. The collapse may have then spread to the west, causing interior members to continue collapsing. The building at this point may have had extensive interior structural failures that then led to the collapse of the overall building, including the cantilever transfer girders along the north elevation, the strong diaphragms at the 5th and 7th floors, and the seat connections between the interior beams and columns at the building perimeter.

8.2.5.2 Recommendations

Certain issues should be explored before final conclusions are reached and additional studies of the performance of WTC 7, and related building performance issues should be conducted. These include the following:

a. Additional data should be collected to confirm the extent of the damage to the south face of the building caused by falling debris.

b. Determination of the specific fuel loads, especially at the lower levels, is important to identify possible fuel supplied to sustain the fires for a substantial duration. Areas of interest include storage rooms, file rooms, spaces with high-density combustible materials, and locations of fuel lines. The control and operation of the emergency power system, including generators and storage tanks, needs to be thoroughly understood. Specifically, the ability of the diesel fuel pumps to continue to operate and send fuel to the upper floors after a fuel line is severed should be confirmed.

c. Modeling and analysis of the interaction between the fires and structural members are important. Specifically, the anticipated temperatures and duration of the fires and the effects of the fires on the structure need to be examined, with an emphasis on the behavior of transfer systems and their connections.

d. Suggested mechanisms for a progressive collapse should be studied and confirmed. How the collapse of an unknown number of gravity columns brought down the whole building must be explained.

e. The role of the axial capacity between the beam-column connection and the relatively strong structural diaphragms may have had in the progressive collapse should be explained.

f. The level of fire resistance and the ratio of capacity-to-demand required for structural members and connections deemed to be critical to the performance of the building should be studied. The collapse of some structural members and connections may be more detrimental to the overall performance of the building than other structural members. The adequacy of current design provisions for members whose failure could result in large-scale collapse should also be studied.

8.2.6 Chapter 6: Bankers Trust Building

8.2.6.1 Observations and Findings

a. An evaluation of the damage patterns revealed several interesting interpretations. The spandrels were sheared by the impactor, between column lines C and D, from the 23rd to the 19th floors. The D-8 column splices failed at the 18th floor and at the 16th floor, but there are no clues to indicate why

CHAPTER 8: *Observations, Findings, and Recommendations*

column splice tension overload occurred at this location. However, unlike the spandrels above, the girder-column connections at column lines C and D failed. Although severed from the column above and below, column D-8 remained suspended from the girders spanning between column lines E and D. These girders developed large vertical and lateral deformations (twisting). The twisting and bending of these girders may have extended the zone of collapse to bays bounded by column lines C and E. If the column splices had not failed at the 16th and 18th floors, it is possible the extent of collapse may have been limited to the single bay in the path of the impactor. This enlarged zone of damage continued until the collapse was arrested on the 8th floor. It is unlikely that dynamic effects caused the damage to column D-8 below the 16th floor; otherwise, the collapse should have progressed all the way to the ground. It is possible that the column splice failures and the resulting large deformations (twisting) of the spandrels caused the remaining portion of column D-8 to lose lateral bracing, and the collapse was not arrested until the energy of the impactor and debris pile was sufficiently diminished to halt the collapse. If this actually accounted for the enlargement of the damage zone, the restraint of the twisting deformations may have prevented the failure of column D-8.

b. Although a considerable amount of debris fell from the upper floors onto the first-floor extension to the north, a two-story deep pile of debris accumulated on the 8th floor. By one estimate, although the debris distributed some of its weight by bridging action, the net effect would have been a 500-percent increase in dead load moment for the supporting beam. Based on the computed results, and in the absence of wind, it appears that the connections would have been able to support more than 500 percent of the estimated dead load moments before any hinging would occur. This may explain why multiple stories of debris came to rest at the 8th floor without incurring additional damage to the structure.

c. Because column D-8 failed below the 16th floor, the beam-to-column moment connection was the single most significant structural feature that helped limit the damage. The portion of the building above the collapsed floors was held in place by frame action of the perimeter. Static elastic analyses of the moment frame show very high stress levels; however, there was a negligible deformation directly above the damaged structure. Furthermore, connections that enable the beams to develop some membrane capacity improve a structure's ability to arrest collapse. The typical floor beam end connections with their A307 bolts were overloaded in direct tension. High-strength bolts would have provided significantly greater tensile ability and possibly held more beams in place through catenary action. Inelastic analyses demonstrate the role of the weaker connections in the response of the structure. Finally, stronger column splices may have made it more difficult for the damaged column to separate from the upper column. Heavier column splices could have allowed the damaged column to function as a hanger and limit the amount of collapsed area, or they could have tended to pull more of the frame down.

8.2.6.2 Recommendations

It is difficult to draw conclusions and more detailed study is required to understand how the collapse was halted. As better descriptions of the structural details become available, the observed patterns of damage may provide useful information in the calibration of numerical simulation tools. Some issues requiring further study are:

a. Whether the observed damage in the column flange, and not at the beam flange, of the moment frames top connection plates is due to high restraint in the welds.

b. Why the bottom flange welded connection has typically failed at the fillet weld to beam interface and not at the fillet weld to seat plate interface.

c. The impact response of various moment-connected details.

FEDERAL EMERGENCY MANAGEMENT AGENCY

d. Whether composite construction would reduce local collapse zones. (There were no shear connectors to provide composite action between the floor beams and slab. Composite construction would have increased the capacity of the members and may have dissipated more of the impact energy; however, it may have also pulled a greater extent of the adjoining regions into the collapse zone.)

e. Whether perimeter rebar in the slabs could improve the structural response by providing catenary action and tensile force resistance in the slabs to reduce local collapse zones.

f. Whether the partial-strength connections permitted members to break away from the structure, thereby limiting the extent of damage. (If the moment connections had been designed for the capacity of the sections [as opposed to fully rigid partial strength based on design load and stiffness requirements], the building performance is likely to have been different.)

g. Whether the collapse zone would have been limited if the spandrels on the 16th, 17th, and 18th floors had not been so grossly distorted through twisting.

8.2.7 Chapter 7: Peripheral Buildings

8.2.7.1 Observations and Findings

a. Steel-frame construction from the 1900s through the 1980s, though different in many details, performed well under significant impact loads by limiting impact damages and progressive collapse to local areas.

b. Heavy unreinforced masonry façades were observed to absorb significant amounts of impact energy in the Verizon and 90 West Street buildings. Heavy masonry façades like those in the Verizon, 90 West Street, or even 130 Cedar Street buildings may also provide an alternative load path for a damaged structure.

c. Older, early-century fireproofing methods of concrete-, brick-, and terra cotta tile-encased steel frames performed well, even after 90+ years, and protected the 90 West Street building from extensive structural damage.

8.2.7.2 Recommendations

The known data and conditions of the perimeter structures after the impact damage should be utilized as a basis for calibration, comparison, and verification of existing software intended to predict such behavior, and for the development of new software for the prediction of the ability of structures to sustain localized and global overload conditions.

8.2.8 Appendix C: Limited Metallurgical Examination

Two structural steel samples from the WTC site were observed to have unusual erosion patterns. One sample is believed to be from WTC 7 and the other from either WTC 1 or WTC 2.

8.2.8.1 Observations and Findings

a. The thinning of the steel occurred by high temperature corrosion due to a combination of oxidation and sulfidation.

b. Heating of the steel into a hot corrosive environment approaching 1,000 °C (1,800 °F) results in the formation of a eutectic mixture of iron, oxygen, and sulfur that liquefied the steel.

c. The sulfidation attack of steel grain boundaries accelerated the corrosion and erosion of the steel.

d. The high concentration of sulfides in the grain boundaries of the corroded regions of the steel occurred due to copper diffusing from the high-strength low-alloy (HSLA) steel combining with iron and sulfur, making both discrete and continuous sulfides in the steel grain boundaries.

8.2.8.2 Recommendations

The severe corrosion and subsequent erosion of Samples 1 and 2 constitute an unusual event. No clear explanation for the source of the sulfur has been identified. The rate of corrosion is also unknown. It is possible that this was the result of long-term heating in the ground following the collapse of the buildings. It is also possible that the phenomenon started prior to collapse and accelerated the weakening of the steel structure. A detailed study into the mechanisms of this phenomenon is needed to determine what risk, if any, is presented to existing steel structures exposed to severe and long-burning fires.

8.3 Building Performance Study Recommendations for Future Study

The BPS Team has developed recommendations for specific issues, based on the study of the performance of the WTC towers and surrounding buildings in response to the impact and fire damage that occurred. These recommendations have a broader scope than the important issue of building concepts and design for mitigating damage from terrorist attacks, and also address the level at which resources should be expended for aircraft security, how the fire protection and structural engineering communities should increase their interaction in building design and construction, possible considerations for improved egress in damaged structures, the public understanding of typical building design capacities, issues related to the study process and future activities, and issues for communities to consider when they are developing emergency response plans that include engineering response.

8.3.1 National Response

Resources should be directed primarily to aviation and other security measures rather than to hardening buildings against airplane impact. The relationship and cooperation between public and private organizations should be evaluated to determine the most effective mechanisms and approaches in the response of the nation to such disasters.

8.3.2 Interaction of Structural Elements and Fire

The existing prescriptive fire resistance rating method (ASTM E119) does not provide sufficient information to determine how long a building component in a structural system can be expected to perform in an actual fire. A method of assessing performance of structural members and connections as part of a structural system in building fires is needed for designers and emergency personnel.

The behavior of the structural system under fire conditions should be considered as an integral part of the structural design. Recommendations are to:

- Develop design tools, including an integrated model that predicts heating conditions produced by the fire, temperature rise of the structural component, and structural response.

- Provide interdisciplinary training in structures and fire protection for both structural engineers and fire protection engineers.

Performance criteria and test methods for fireproofing materials relative to their durability, adhesion, and cohesion when exposed to abrasion, shock, vibration, rapid temperature rise, and high-temperature exposures need further study.

8.3.3 Interaction of Professions in Design

The structural, fire protection, mechanical, architectural, blast, explosion, earthquake, and wind engineering communities need to work together to develop guidance for vulnerability assessment, retrofit, and the design of concrete and steel structures to mitigate or reduce the probability of progressive collapse under single- and multiple-hazard scenarios.

An improved level of interaction between structural and fire protection engineers is encouraged. Recommendations are to:

- Consider behavior of the structural system under fire as an integral part of the design process.

- Provide cross-training of fire protection and structural engineers in the performance of structures and building fires.

8.3.4 Fire Protection Engineering Discipline

The continued development of a system for performance-based design is encouraged. Recommendations are to:

- Improve the existing models that simulate fire and spread in structures, as well as the impact of fire and smoke on structures and people.

- Improve the database on material burning behavior.

8.3.5 Building Evacuation

The following topics were not explicitly examined during this study, but are recognized as important aspects of designing buildings for impact and fire events. Recommendations for further study are to:

- Perform an analysis of occupant behavior during evacuation of the buildings at WTC to improve the design of fire alarm and egress systems in high-rise buildings.

- Perform an analysis of the design basis of evacuation systems in high-rise buildings to assess the adequacy of the current design practice, which relies on phased evacuation.

- Evaluate the use of elevators as part of the means of egress for mobility-impaired people as well as the general building population for the evacuation of high-rise buildings. In addition, the use of elevators for access by emergency personnel needs to be evaluated.

8.3.6 Emergency Personnel

One of the most serious dangers firefighters and other emergency responders face is partial or total collapse of buildings. Recommended steps to provide better protection to emergency personnel are to:

- Have fire protection and structural engineers assist emergency personnel in developing pre-plans for buildings and structures to include more detailed assessments of hazards and response of structural elements and performance of buildings during fires, including identification of critical structural elements.

- Develop training materials and courses for emergency personnel with regard to the effects of fire on steel.

- Review collaboration efforts between the emergency personnel and engineering professions so that engineers may assist emergency personnel in assessments during an incident.

8.3.7 Education of Stakeholders

Stakeholders (e.g., owners, operators, tenants, authorities, designers) should be further educated about building codes, the minimum design loads typically addressed for building design, and the extreme events that are not addressed by building codes. Should stakeholders desire to address events not included in the building codes, they should understand the process of developing and implementing strategies to mitigate damage from extreme events.

Stakeholders should also be educated about the expected performance of their building when renovations, or changes in use or occupancy, occur and the building is subjected to different floor or fire loads. For instance, if the occupancy in a building changes to one with a higher fire hazard, stakeholders should have the fire protection systems reviewed to ensure there is adequate fire protection. Or, if the structural load is increased with a new occupancy, the structural support system should be reviewed to ensure it can carry the new load.

8.3.8 Study Process

This report benefited from a tremendous amount of professional volunteerism in response to this unprecedented national disaster. Improvements can be made that would aid the process for any future efforts. Recommendations are to:

- Provide resources that are proportional to the required level of effort.

- Provide better access to data, including building information, interviews, samples, site photos, and documentation.

8.3.9 Archival Information

Archival information has been collected and provides the groundwork for continued study. It is recommended that a coordinated effort for the preservation of this and other relevant information be undertaken by a responsible organization or agency, capable of maintaining and managing such information, This effort would include:

- cataloging all photographic data collected to date

- enhancing video data collected for both quality and timeline

- conducting interviews with building occupants, witnesses, rescue workers, and any others who may provide valuable information

- initiating public requests for information

8.3.10 SEAoNY Structural Engineering Emergency Response Plan

As with any first-time event, difficulties were encountered at the beginning of the relationship between the volunteer engineering community and the local government agencies. Lessons learned in hindsight can be valuable to other engineering and professional organizations throughout the country. Appendix F presents recommendations that can be used as a basis for the development of other, similar plans.

James Milke
Venkatesh Kodur
Christopher Marrion

A Overview of Fire Protection in Buildings

A.1. Introduction

This appendix presents background information on the fire and life safety aspects of buildings for the interested reader. This review of fire behavior outlines burning characteristics of materials as well as the effect of building characteristics on the temperatures experienced. The description of the effect of fire exposure on steel and concrete structural members is intended to improve understanding of how these structural members respond when heated and also what measures are commonly used to limit temperature rise in structural members. Finally, a brief discussion on evacuation behavior in high-rise buildings is included to provide some context to the comments made in the report concerning the design of the means of egress and the evacuation process in WTC 1 and WTC 2.

A.2 Fire Behavior

Important aspects of fire behavior in the affected buildings involves the following issues:

- burning behavior of materials, including mass loss and energy release rates

- stages of fire development

- behavior of fully developed fires, including the role of ventilation, temperature development, and duration

A.2.1 Burning Behavior of Materials

Once a material is ignited, a fire spreads across the fuel object until it becomes fully involved. The spread at which flame travels over the surface of the material is dependent on the fuel composition, orientation, surface to mass ratio, incident heat, and air supply. Given sufficient air, the energy released from a fire is dictated by the incident heat on the fuel and the fuel characteristics, most notably the heat of combustion and latent heat of vaporization. The relationship of these parameters to the energy release rate is given by:

$$\dot{Q}'' = \frac{\dot{q}''}{L_v} \Delta H_c \qquad (A\text{-}1)$$

FEDERAL EMERGENCY MANAGEMENT AGENCY

where:

$\dot{Q}''$ = energy release rate per unit surface area of fuel

$\dot{q}''$ = incident heat per unit surface area of fuel (i.e., heat flux)

L_v = latent heat of vaporization

ΔH_c = heat of combustion

The effective heat of combustion for a mixture of wood and plastics is on the order of 16 kJ/g. For fully developed fires, the radiant heat flux is approximately 150 to 200 kW/m². The latent heat of vaporization for a range of wood and plastics is 5 to 8 kJ/g. Thus, the mass burning rate per unit surface area in typical office building fires ranges from 20 to 40 g/m²-s and the associated energy release rate per unit surface area ranges from 320 to 640 kW/m².

In typical fires, as the fire grows in size, the energy release rate increases to a peak value as depicted in Figure A-1. The increase in the heat release rate with time depends on the fuel characteristics, incident heat, and available air supply. Sample curves for alternate materials, described in the fire protection literature as "slow," "medium," and "fast" growth rate fires, are illustrated in Figure A-2.

At some point, the heat release rate of the fire will become limited by either the amount of fuel or the amount of oxygen that is available; this is referred to as the peak heat release rate. Peak heat release rate data can be obtained through experimental testing and is available for many types of materials and fuels. Table A.1 includes a list of selected common items and their associated peak heat release rates.

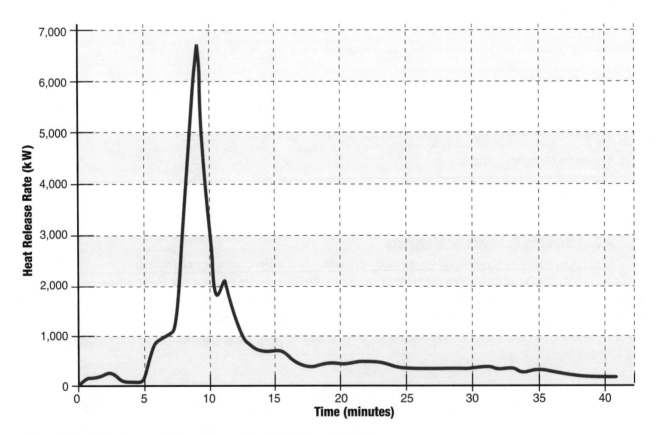

Figure A-1 Heat release rate for office module (Madrzykowski 1996).

APPENDIX A: *Overview of Fire Protection in Buildings*

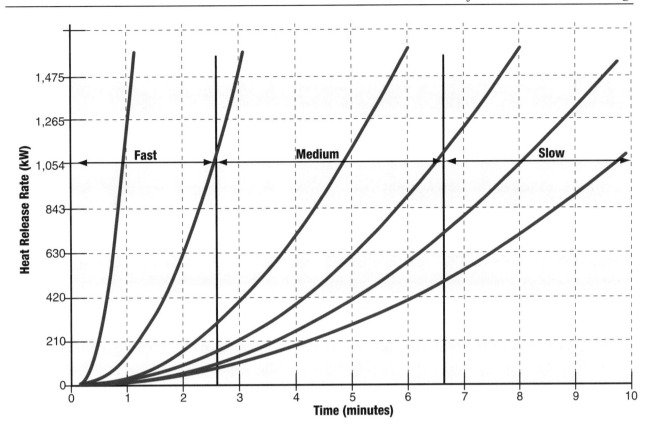

Figure A-2 Fire growth rates (from SFPE Handbook of Fire Protection Engineering).

Table A.1 Peak Heat Release Rates of Various Materials (NFPA 92B and NFPA 72)

Item	Heat Release Rate
Crumpled brown lunch bag, 6 g	1.2 kW
Folded double-sheet newspaper, 22 g	4 kW
Crumpled double sheet newspaper, 22 g	17 kW
Medium wastebasket with milk cartons	100 kW
Plastic trash bag with cellulosic material (1.2–14 kg)	120–350 kW
Upholstered chair with polyurethane foam	350 kW
Christmas tree, dry	500-650 kW
Latex foam mattress (heat at room door)	1,200 kW
Furnished living room	4,000–8,000 kW

After a fire has reached its peak heat release rate, it will decline after some period of time. At this point, most of the available fuel has typically been burned and the fire will slowly decrease in size. The length of the decay phase depends on what type of fuel is available, how complete was the combustion of the fuel, how much oxygen is present in the compartment, and whether any type of suppression is occurring. The burning rate of liquid fuels is on the order of 50 g/m^2-s, with an associated energy release rate per unit surface area of approximately 2,000 kW/m^2. The burning rate per unit area of information is useful to estimate the duration of a fire involving a particular fuel spread over a specified area.

A.2.2 Stages of Fire Development

Generally, fires are initiated within a single fuel object. The smoke produced from the burning object is transported by a smoke plume and collects in the upper portion of the space as a layer. The smoke plume also transports the heat produced by the fire into the smoke layer, causing the smoke layer to increase in depth and also temperature. This smoke layer radiates energy back to unburned fuels in the space, causing them to increase in temperature.

Fire spreads to other objects either by radiation from the flames attached to the originally burning item or from the smoke layer. As other objects ignite, the temperature of the smoke layer increases further, radiating more heat to other objects. In small compartments, the unburned objects may ignite nearly simultaneously. This situation is referred to as "flashover." In large compartments, it is more likely that objects will ignite sequentially. The sequence of the ignitions depends on the fuel arrangement, and composition and ventilation available to support combustion of available fuels.

A.2.3 Behavior of Fully Developed Fires

A fully-developed fire is one that reaches a steady state burning stage, where the mass loss rate is relatively constant during that period. The equilibrium situation may occur as a result of a limited ventilation supply (in ventilation controlled fires) or due to characteristics of the fuel (fuel-controlled fires).

If the rate of mass burning based on the incident heat flux and fuel characteristics (see Section A.2.1) exceeds the amount that can be supported by the available air supply, the burning becomes ventilation controlled. Otherwise, the fire is referred to as being fuel controlled. The ventilation air for the fires may be supplied from openings to the room, such as open windows or doors, or other sources such as HVAC systems.

Given that the heat released per unit of oxygen is a relatively constant value of 13.1 kJ/g for common fuels, the air supply required to support fires of a particular heat release rate can be determined. For every 1 MW of heat release rate, 76 g/s of oxygen is consumed. Considering that air is 21 percent oxygen, this flow of oxygen requires a flow of 0.24 m^3/s (500 cfm) of ambient air. In the case of WTC 1 and WTC 2, for a 3-GW fire, a flow of 1,500,000 cfm of air was required to support that fire. That airflow would have been supplied via openings in the exterior wall and the shaft walls.

Most of the research on fully-developed fires has been conducted in relatively small spaces with near-square floor plans. In such cases, the conditions (temperature of the smoke and incident heat on the enclosure) are relatively uniform throughout the upper portion of the space. However, Thomas and Bennetts (1999) have documented differences in that behavior for ventilation controlled fires in long, thin spaces or in large areas. In such cases, the burning occurs in the fuel nearest to the supply source of air. Temperatures are observed to be greatest nearest to the supply source of air.

In large or complex buildings, the incident flux on the structural elements is expected to vary over the entire space of fire involvement. A range of developing numerical models have the ability to compute the variation of the fire imposed heat flux on a 3-dimensional grid. The Fire Dynamics Simulator from the

National Institute of Standards and Technology (NIST) is an example of such a model that has the promise of developing into a tool that could be used to estimate the variation in incident heat flux on structural elements over a large space of fire involvement.

A.3 Structural Response to Fire

A.3.1 Effect of Fire on Steel

A.3.1.1 Introduction

Fire resistance is defined as the property of a building assembly to withstand fire, or give protection from it (ASTM 2001a). Included in the definition of fire resistance are two issues. The first issue is the ability of a building assembly to maintain its structural integrity and stability despite exposure to fire. Secondly, for some assemblies such as walls and floor-ceiling assemblies, fire resistance also involves serving as a barrier to fire spread.

Fire resistance is commonly assessed by subjecting a prototype assembly to a standard test. Results from the test are reported in terms of a fire resistance rating, in units of hours, based on the time duration of the test that the building assembly continues to satisfy the acceptance criteria in the test.

Fire resistance rating requirements for different building components are specified in building codes. These ratings depend on the type of occupancy, number of stories, and floor area. Because the standard test is intended to be a comparative test and is not intended to predict actual performance, the hourly fire resistance ratings acquired in the tests should not be misconstrued to indicate a specific duration that a building assembly will withstand collapse in an actual fire.

Generally, the fire resistance rating of a structural member is a function of:

- applied structural load intensity,
- member type (e.g., column, beam, wall),
- member dimensions and boundary end conditions,
- incident heat flux from the fire on the member or assembly,
- type of construction material (e.g., concrete, steel, wood), and
- effect of temperature rise within the structural member on the relevant properties of the member.

The fire performance of a structural member depends on the thermal and mechanical properties of the materials of which the building component is composed. As a result of the increase in temperature caused by the fire exposure, the strength of steel decreases along with its ability to resist deformation, represented by the modulus of elasticity. In addition, deformations and other property changes occur in the materials under prolonged exposures. Likewise, concrete is affected by exposure to fire and loses strength and stiffness with increasing temperature. In addition, concrete may spall, resulting in a loss of concrete material in the assembly. Spalling is most likely in rapid-growth fires, such as may have occurred in WTC 1 and WTC 2.

The performance of fire-exposed structural members can be predicted by structural mechanics analysis methods, comparable to those applied in ambient temperature design, except that the induced deformations and property changes need to be taken into consideration.

Beams and trusses may react differently to severe fire exposures, depending on the end conditions and fabrication. Unconnected members may collapse when the stresses from applied loads exceed the available strength for beams and trusses. In the case of connected members, significant deflections may occur as a result of reduced elastic modulus, but structural integrity is preserved as a result of catenary action.

APPENDIX A: *Overview of Fire Protection in Buildings*

In the case of slender columns, the susceptibility for buckling increases with a decrease in the modulus of elasticity. Where connections of floor framing to columns fail, either at the ends or intermediate locations, column slenderness is increased, thereby increasing the susceptibility of a column to buckling.

Steels most often used in building design and construction are either hot-rolled or cold-drawn. Their strength depends mainly on their carbon content, though some structural steels derive a portion of their strength from a process of heat treatment known as quenching and tempering (e.g., ASTM A913 for rolled shapes and ASTM A325 and A490 for bolts).

A.3.1.2 Evaluating Fire Resistance

Performance Criteria

Building code requirements for structural fire protection are based on laboratory tests conducted in accordance with ASTM E119, Standard Test Methods for Fire Tests of Building Construction and Materials (2000). In these tests, building assemblies, such as floor-ceilings, columns, and walls are exposed to heating conditions created in a furnace, following a specified time-temperature curve. In Figure A-3, time-temperature curves are presented for the standard fire exposure specified in ASTM E119, the standard hydrocarbon exposure in ASTM E1529, and a real building fire. As can be seen, each is somewhat different.

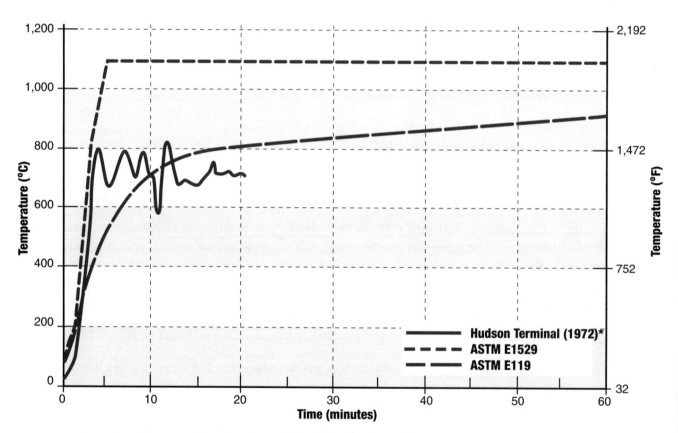

*Hudson Terminal experiment conducted with normal office fuel load (6 psf) (DeCicco, et al. 1972)

Figure A-3 Comparison of exposure temperatures in standard tests.

There are three performance criteria in the standard ASTM E119 test method. These are related to loadbearing capacity, insulation, and integrity:

1. **Loadbearing capacity:** For loadbearing assemblies, the test specimen shall not collapse in such a way that it no longer performs the loadbearing function for which it was constructed.

2. **Insulation:** For assemblies such as floors-ceilings and walls that have the function of separating two parts of a building,

    a. the average temperature rise at the unexposed face of the specimen shall not exceed 139 °C (282 °F), and

    b. the maximum temperature rise at the unexposed face of the specimen shall not exceed 181 °C (358 °F).

3. **Integrity:** For assemblies such as walls, floors, and roofs, the formation of openings through which flames or hot gases can pass shall not occur. Loss of integrity is deemed to have occurred when a specified cotton wool pad applied to the unexposed face is ignited.

Tests are conducted on prototype designs. The fire-resistance rating applies to replicates of the tested assembly, with limited changes permitted. Rules, guidelines, and correlations are available to assess the impact of changes or to develop acceptable variations to the design (ASCE/SFPE 1999).

ASTM E119

The ASTM E119 test is a comparative test and is not intended to be predictive. The test fire exposure, while recognized as severe, is not representative of all fires. Heat transfer conditions associated with the exposing fire are different than those in actual fires. Further, the test is not a full-scale test, with no attempt to scale the response of the test specimen to actual size building assemblies. Although the test requires that floor-ceiling specimens be representative of actual building construction, achieving this in a 14-foot by 17-foot test specimen is difficult. Consequently, ASTM E119 is principally a thermal test, not a structural test, even though the test floor is loaded. Loading of floors and roofs is done to see if the fireproofing material will be dislodged by deflection and buckling of the steel during a fire.

Further, several factors are not applied in this test method, including structural framing continuity, member interaction, restraint conditions, and applied load intensity. The test only evaluates the performance of a building assembly, such as a wall or floor-ceiling assembly. The test does not consider the interaction between adjacent assemblies or the behavior of the structural frame. In "real" buildings, beam/girder/column connections range from simple shear to full moment connections and framing member size and geometry vary significantly, depending on the structural system and building size and layout.

In the Underwriters Laboratories, Inc. (UL) version of the ASTM E119 test, UL 263, the beams are placed on shelf angles and steel wedges are driven by sledgehammers between the end of the beam and the heavy massive steel and concrete furnace frame. This is referred to as a "restrained beam," and the fire test results are published in Volume 1 of the UL Fire Resistance Directory, which is the major reference used by architects and engineers to select designs that meet the building code requirements for fire resistance ratings. The UL Fire Resistance Directory also publishes unrestrained fire resistance ratings based on critical temperature rise in the steel member as discussed in Section A.3.1.6. In spite of the ASTM E119 test limitations relative to the structural conditions that exist in real buildings, the fire test is conservative to the point that more fire protection material is required than has been demonstrated necessary in large scale fire tests conducted and reported in the international fire research literature.

APPENDIX A: *Overview of Fire Protection in Buildings*

There has been much interest in revising the ASTM E119 Standard Fire Test. Arguments are posed that the fire exposure is too severe, while others suggest that the fire exposure is not severe enough. A good compromise is a performance oriented analysis using design fire curves for very specific occupancies and building geometry while still permitting the use of ASTM E119 for general applications.

For most of the 1900s, there was a single U.S. standard time-temperature curve described by ASTM E119. Most of the world adopted that curve or one similar in running the test furnaces.

In 1928, Ingberg of the National Bureau of Standards published a paper on the severity of fire (Ingberg 1928) in which he equated the gross combustible fuel load (combustible content in mass per unit area) to the potential fire exposure in terms of duration of exposure to a fire following the standard (ASTM E119) fire curve. Although subsequent research has shown the simple relationship proposed by Ingberg holds only in limited cases where the fire ventilation is the same as that present in his test series, his equation is still widely published in texts and used as the basis of regulation.

In the 1950s and 1960s, it was demonstrated that, for severe, fully involved fires, the intensity and duration of burning within compartments and other enclosures were also functions of the availability of air for combustion, commonly referred to as ventilation and normally coming from openings such as doors and broken windows or from forced ventilation from the HVAC system.

In Sweden, an extensive family of fire curves has been developed, by test, for fully involved (i.e., post flashover) fires as a combined function of fuel load and ventilation (Magnusson and Thelandersson 1970). The published curves have peak temperatures of 600–1,100 °C (1,100–2,000 °F).

Most recently, Ian Thomas in Australia has demonstrated with reduced scale models that the combustion process in facilities where there is a depth from the vent opening (e.g., broken windows) to the actual fuel can produce conditions where a large portion of the vaporized fuel actually burns at a point removed from the location of the solid fuel (combustible material) source. Thomas' experiments used fully involved spaces where the depth from the vent opening was at least twice the width of the test space. In these experiments, the air supply drawn into the test space by the fire was insufficient to burn all of the available fuel. Fuel once vaporized was transported to the openings and burned there, producing an unexpectedly high heat flux on the elements at and near the vent opening. The importance of Thomas' work is that it demonstrates the fact that, in many fires, the reality is that the fire exposing the structural elements is not necessarily a constant in either time or space.

Fortunately, there are now advanced numerical models capable of describing the fire caused environment in detail.

ASTM E1529 and UL 1709: The Hydrocarbon Pool Curves

In the late 1980s, as a result of failures of fireproofed steel members exposed to petroleum spill fires, the petroleum industry felt a need to develop a new test curve. The curve developed was designed to apply a sudden and intense shock, typified by a large hydrocarbon pool fire either burning in the open or in some other situation where there was no significant restraint to the flow of combustion air to the burning pool fire. ASTM E1529 was developed to answer this need. The objective of this ASTM test is to almost instantaneously impose 158 kW/m^2 (50,000 Btu/ft^2-hr) on the element under test. Additionally a similar but somewhat more severe test procedure has been developed by Underwriters Laboratories and published as their standard UL 1709. The UL test is designed to impose 200 kW/m^2 (65,000 Btu/ft^2-hr) on the test element. This unusual difference in the ASTM and UL standards reflects a technical difference of opinion between the two organizations. The tests are often quoted as a time-temperature curve quickly reaching and maintaining a test furnace temperature of 1,093 °C (2,000 °F) in the case of the ASTM standard and 1,143

°C (2,089 °F) at UL. The hydrocarbon time-temperature curve is, however, actually a test-specific item and can vary some from test apparatus to test apparatus.

The ASTM E119 curve was derived from experiments and is empirically based; however, ASTM E1529 exposure is based on judgment, experience, and a database of experiments concerning the measurement of the temperatures involved in large hydrocarbon fires. The incident flux approximates the incident flux on a member completely bathed in the flame from a large free-burning pool fire. Although both of the ASTM curves are useful in conducting tests of fireproofed building elements as pre-installment tests, they are not predictions of the intensity of actual fires and are often not appropriate as an input to models or other computations seeking to assess a fire hazard for a building.

A prime impact of the high flux "shock" exposure is to test the capability of the fireproofing to survive such exposure. In addition, such thermal shock could induce spalling in concrete systems.

Comparison between ASTM E119, ASTM E1529, and UL 1709 is further complicated by instrumentation differences in the two "hydrocarbon fire" tests and that used in the ASTM E119 test. In particular, different thermocouple installations are used to control and record furnace temperatures in the respective tests. In the ASTM E119 test, the thermocouples are contained within a protective capped steel pipe, resulting in a time delay between the actual and recorded furnace temperatures. In the hydrocarbon tests, the thermocouples are bare, thereby providing a more timely indication of the actual gas temperature. The lag in ASTM E119 is most pronounced at the start of the test. Figure A-3 provides a plot of the two standard curves with an additional curve of the approximate actual temperature (if measured with bare thermocouples) in an ASTM E119 furnace test. Most of the tests to date have been conducted using the UL 1709 curve. Many tested items show a significantly shorter time to failure using the UL 1709 procedure as compared to the ASTM E119 procedure.

A.3.1.3 Response of High-rise, Steel-frame Buildings in Previous Fire Incidents

In recent years, three notable fires have occurred in steel frame buildings, though none involved the total floor area as in WTC 1 and WTC 2. However, prior to September 11, 2001, no protected steel frame buildings had been known to collapse due to fire. These previous three fire incidents include the following:

- 1st Interstate Bank Building, Los Angeles, May 4-5, 1988

- Broadgate Phase 8, UK, 1990

- One Meridian Plaza, Philadelphia, February 23-24, 1991

The steel in the 1st Interstate Bank Building and One Meridian Plaza was protected with spray-applied protection. Because the fire occurred at the Broadgate complex while it was under construction, the steel beams had not yet been protected. The fire durations of the three incidents are indicated in Table A.2. The durations noted in the table refer to the overall duration of the incident. The fire duration in a particular area of the building was likely less than that noted.

In the case of the fire at One Meridian Plaza, the fire burned uncontrolled for the first 11 hours and lasted 19 hours. Contents from nine floors were completely consumed in the fire. In addition to these experiences in fire incidents, as a result of the Broadgate fire, British Steel and the Building Research Establishment performed a series of six experiments at Cordington in the mid-1990s to investigate the behavior of steel frame buildings. These experiments were conducted in a simulated, eight-story building. Secondary steel beams were not protected. Despite the temperature of the steel beam reaching 800–900 °C (1,500–1,700 °F) in three tests (well above the traditionally assumed critical temperature of 600 °C [1,100 °F]), no collapse was observed in any of the six experiments.

Table A.2 Fire Duration in Previous Fire Incidents in Steel-frame Buildings

Building	Date	Fire Duration (hours)
1st Interstate Bank Building	May 4-5, 1988	3.5
Broadgate Phase 8	1990	4.5
One Meridian Plaza	February 23-24, 1991	19 (11 uncontrolled)

One important aspect of these previous incidents is that the columns remained intact and sustained their load carrying ability throughout the fire incidents (though there was no structural damage caused by impacts). Throughout the fire in One Meridian Plaza, horizontal forces were exerted on the columns by the girders and despite the 24- to 36-inch deflections of the girders, floor beams, and concrete and steel deck floor slabs, the columns continued to stabilize the building throughout the fire and for several years after the fire.

Questions have been raised about the comparison of the structural performance of the WTC 1 and WTC 2 and the Empire State Building. In the case of the Empire State Building:

1. The impacting aircraft was a U. S. Army Air Force B-25 bomber weighing 12 tons with a fuel capacity of 975 gallons, which, at the time of the crash, was traveling at a speed estimated to be 250 mph;

2. Crash damage to structural steel was confined to three steel beams. One exterior wall column withstood the direct impact without visible effect;

3. Exterior walls are ornamental cast aluminum panels under windows with steel trim backed by 8 inches of brick. The walls at columns are 8 inches of limestone backed by 8 inches of brick supported on steel framing; and

4. The floors above the Saturday morning plane crash were largely vacant and unoccupied, so the fire load was minimal and perhaps close to zero. Fire was confined to a portion of two floors. Because the building had few occupants at the time of the crash, the fire department could concentrate on controlling and extinguishing the fire.

A.3.1.4 Properties of Steel

The principal thermal properties that influence the temperature rise and distribution in a member are its thermal conductivity, specific heat, and density. The temperature-dependence of the thermal conductivity and specific heat for steel are depicted in Figure A-4.

The mechanical properties that affect the fire performance of structural members are strength, modulus of elasticity, coefficient of thermal expansion, and creep of the component materials at elevated temperatures. Information on the thermal and mechanical properties at elevated temperatures for various types of steel is available in the literature (Lie 1992, Milke 1995, Kodur and Harmathy 2002).

References to the tensile or compressive strength of steel relate either to the yield strength or ultimate strength. Figure A-5 shows the stress-strain curves for a structural steel (ASTM A36) at room temperature and elevated temperatures. As indicated in the figure, the yield and ultimate strength decrease with temperature as does the modulus of elasticity. Figure A-6 shows the variation of strength with temperature (ratio of strength at elevated temperature to that at room temperature) for hot rolled steel such as A36. As indicated in the figure, if the steel attains a temperature of 550 °C (1,022 °F), the remaining strength is approximately half of the value at ambient temperature.

APPENDIX A: *Overview of Fire Protection in Buildings*

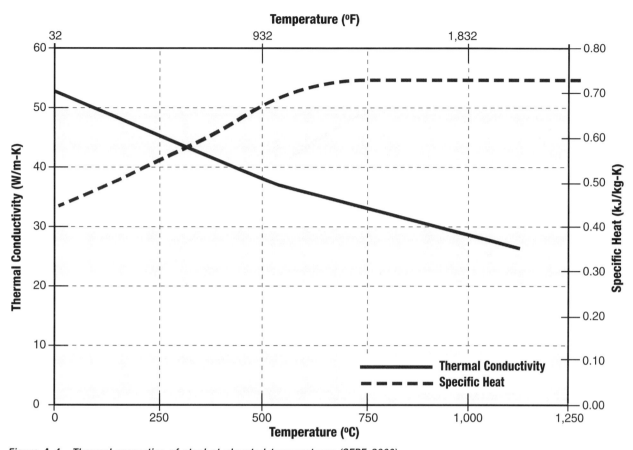

Figure A-4 Thermal properties of steel at elevated temperatures (SFPE 2000).

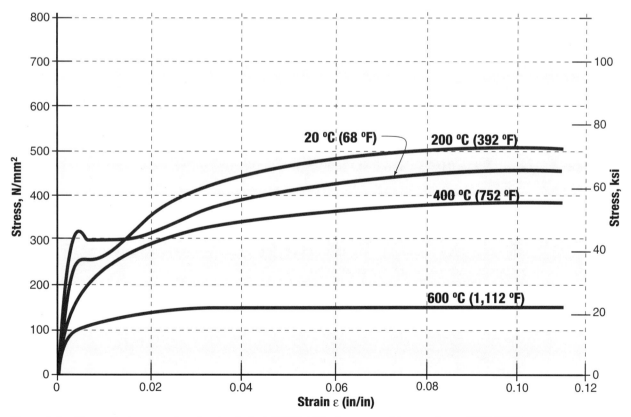

Figure A-5 Stress-strain curves for structural steel (ASTM A36) at a range of temperatures (SFPE 2000).

APPENDIX A: *Overview of Fire Protection in Buildings*

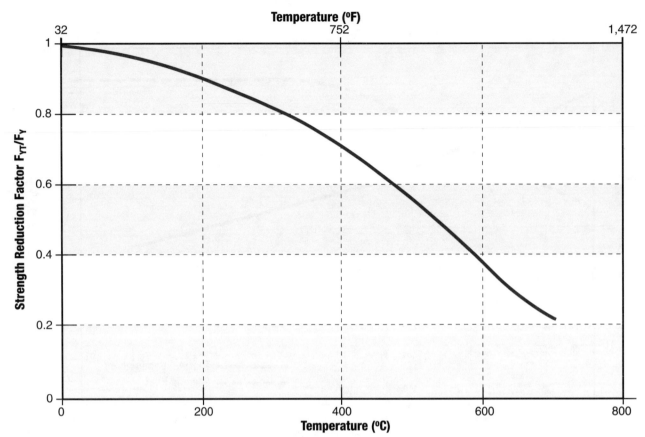

Figure A-6 Strength of steel at elevated temperatures (Lie 1992).

The modulus of elasticity, E_0, is about 210×10^3 MPa for a variety of common steels at room temperature. The variation of the modulus of elasticity with temperature for structural steels and steel reinforcing bars is presented in Figure A-7. As in the case of strength, if the steel attains a temperature of 550 °C (1,022 °F), the modulus of elasticity is reduced to approximately half of the value at ambient temperature.

Figure A-8 shows the variation of yield strength of light gauge steel at elevated temperatures, corresponding to 0.5 percent, 1.5 percent, and 2 percent strains based on the relationships in Gerlich (1995), Makelainen and Miller (1983), and BSI (2000).

In addition to the changes in the properties with increasing temperature, steel expands with increasing temperature. The coefficient of thermal expansion for structural steel is approximately 11×10^{-6} mm/mm-°C. Consequently, an unrestrained, 20-meter-long steel member that experiences a temperature increase of 500 °C (1,022 °F) will expand approximately 110 mm. WTC 5 had many buckled girders and beams on the burned-out fire floors where the expansion was restrained.

An approximate melting point for steel is 1,400 °C (2,500 °F); however, the melting temperature for a particular steel component varies with the steel alloy used.

APPENDIX A: *Overview of Fire Protection in Buildings*

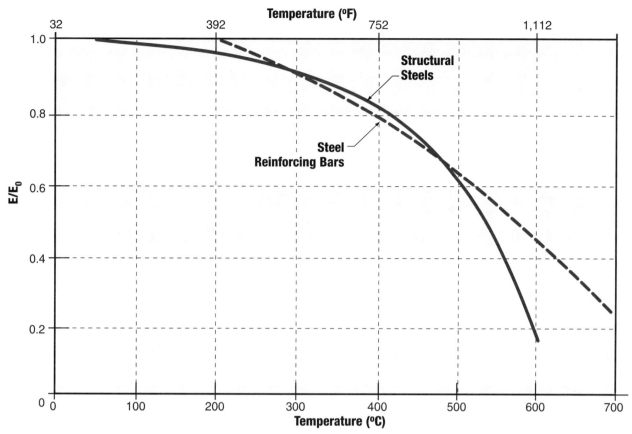

Figure A-7 Modulus of elasticity at elevated temperatures for structural steels and steel reinforcing bars (SFPE 2000).

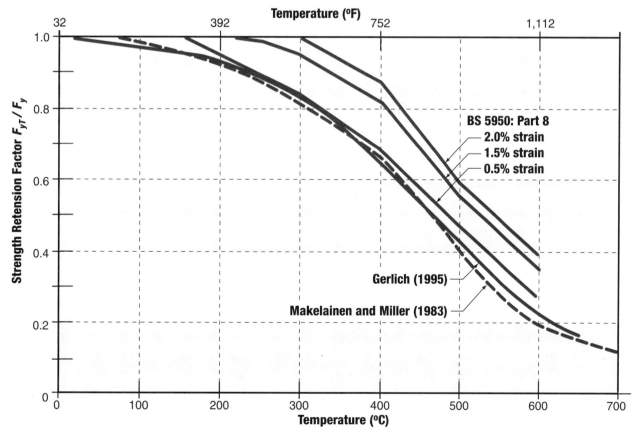

Figure A-8 Reduction of the yield strength of cold-formed light-gauge steel at elevated temperatures.

A.3.1.5 Fire Protection Techniques for Steel

Given the significant reduction in the mechanical properties of steel at temperatures on the order of 540 °C (1,000 °F), isolated and unprotected steel members subjected to the standard test heating environment are only able to maintain their structural integrity for 10 to 20 minutes, depending on the mass and size of the structural member. Unprotected open web steel joists supporting concrete floors in the ASTM E119 fire test have been tested and collapse in 7 minutes (Wang and Kodur 2000).

Isolated and unprotected steel box columns 8 inches x 6 1/2 inches formed using 1/4-inch plate and channels in an ASTM E119 fire test collapse in about 14 minutes (Kodur and Lie 1995). Consequently, measures are taken to protect loadbearing, steel structural members where the members are part of fire resistant assemblies. A variety of methods are available to limit the temperature rise of steel structural members, including the insulation method and the capacitive method.

Insulation Method: The insulation method consists of attaching insulating spray-applied materials, board materials, or blankets to the external surface of the steel member. A variety of insulating materials have been used following this method of protection, including mineral-fiber or cementitious spray-applied materials, gypsum wallboard, asbestos, intumescent coatings, Portland cement concrete, Portland cement plaster, ceramic tiles, and masonry materials. The insulation may be sprayed directly onto the member being protected, such as is commonly done for steel columns, beams, or open web steel joists. The spray-applied mineral fiber, fire resistive coating is a factory mixed product consisting of manufactured inorganic fibers, proprietary cement-type binders, and other additives in low concentrations to promote wetting, set, and dust control. Air setting, hydraulic setting, and ceramic setting binders can be used in varying quantities and combinations or singly, depending on the particular application.

Alternatively, the insulation may be used to form a "membrane" around the structural member, in which case a fire resistive barrier is placed between a potential fire source and the steel member. An example of membrane protection is a suspended ceiling positioned below open web steel joists. (In order for a suspended ceiling assembly to perform effectively as a membrane form of protection, it must remain in place despite the fire exposure. Only some suspended ceiling assemblies have this capability.)

In most of the WTC complex buildings and tall buildings built over the last 50 years, the preferred method has been spray-applied mineral fiber or cementitious materials. Of these 50 years, for the first 20 years the product contained asbestos and for the last 30 years it has been asbestos free. The WTC 1, 2, and 7 incidents are the first known collapses of fire resisting steel frame buildings protected with this type of fireproofing material. Occasionally, a portion of the steel is protected with a spray or trowel applied plaster or Portland cement (e.g., Gunite or shotcrete).

Capacitive Method: The capacitive heat sink method is based on the principle of using the heat capacity of a protective material to absorb heat. In this case, the supplementing material absorbs the heat as it enters the steel and acts as a heat sink. Common examples include concrete filled hollow steel columns and water filled hollow steel columns (Kodur and Lie 1995). In addition, a concrete floor slab may act as a heat sink to reduce the temperature of a supporting beam or open web steel joist.

A.3.1.6 Temperature Rise in Steel

In building materials such as steel, a critical temperature is often referenced at which the integrity of fully-loaded structural members becomes questionable. The critical temperature for steel members varies with the type of steel structural member (e.g., beams, columns, bar joists, or reinforcing steel). North American Test Standards (e.g., ASTM E119) assume a critical temperature of 538 °C (1,000 °F) for structural steel columns. The critical temperatures for columns and other steel structural elements are given in Table A.3. The critical temperature is defined as approximately the temperature where the steel has lost approximately 50 percent of its yield strength from that at room temperature. In an actual structure, the actual impact of such heating of the steel will also depend on the actual imposed load, member end restraint (axial and rotational), and other factors as discussed in Section A.3.1.7.

Table A.3 Critical Temperatures for Various Types of Steel

Steel	Temperature
Columns	538 °C (1,000 °F)
Beams	593 °C (1,100 °F)
Open Web Steel Joists	593 °C (1,100 °F)
Reinforcing Steel	593 °C (1,100 °F)
Prestressing Steel	426 °C (800 °F)

To limit the loss of strength and stiffness, external fire protection is provided to the steel structural members to satisfy required fire resistance ratings. This is usually achieved by fire protecting the steel members to keep the temperature of the steel, in case of a fire, from reaching a critical limit. Traditionally, the amount of fire protection needed is based on the results of standard fire resistance tests.

The temperature attained in a fire-exposed steel member depends on the fire exposure, characteristics of the protection provided, and the size and mass of the steel. For steel members protected with direct-applied insulating materials, the role of the insulating materials is strongly dependent on their thermal conductivity and thickness.

The role of the fire exposure and size and mass of the steel can be demonstrated by analyzing the temperature rise in two protected steel columns with two different fire exposures. For this comparative analysis, the fire exposure associated with two standard fire resistance tests is selected, ASTM E119 and UL 1709. The following two column sizes are selected for this comparative analysis:

- W14X193

- steel box column, 36 inches x 16 inches, with a wall thickness of 7/8 inch for the 36-inch-wide side and 15/16 inch for the 16-inch-wide side

In the first analysis, the steel columns are considered to be unprotected. The results of the analysis are presented in Figure A-9. In the second analysis, 1 inch of a spray-applied, mineral fiber insulation material was assumed to be present (the thermal conductivity of the insulation material was assumed to be 0.116 W/m-K). The results of this analysis are presented in Figure A-10.

In both analyses, the resulting steel column temperatures follow expected trends. The more massive column (the tube) experiences less temperature rise for the same fire exposure than the lighter column (the W14x193). The unprotected columns reach critical temperatures exposed to ASTM E119 condition in 15 to 18 minutes. For the more severe UL 1709 exposure, the unprotected columns reach critical temperatures in 6 to 7 minutes. In contrast, the temperature of the protected columns after 2 hours of exposure to the ASTM E119 conditions is 240 °C (464 °F) for the tube, while the temperature of the W14x193 is 330 °C (626 °F). For the more severe fire exposure associated with UL 1709, the temperature of the steel columns after 2 hours is 60–80 °C (140–176 °F) greater than for each of the steel columns exposed to the ASTM E119 conditions.

APPENDIX A: *Overview of Fire Protection in Buildings*

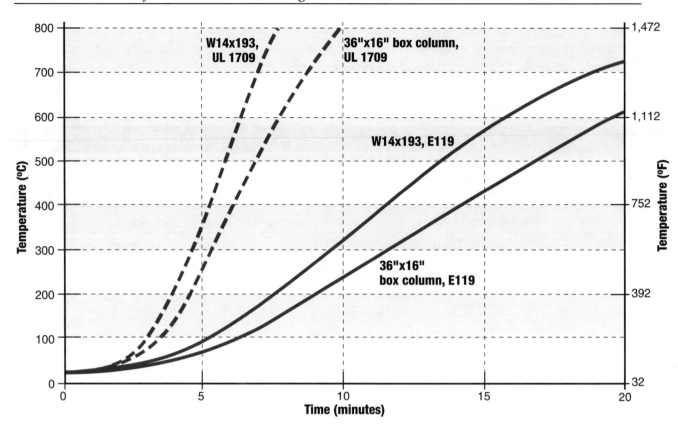

Figure A-9 *Steel temperature rise due to fire exposure for unprotected steel column.*

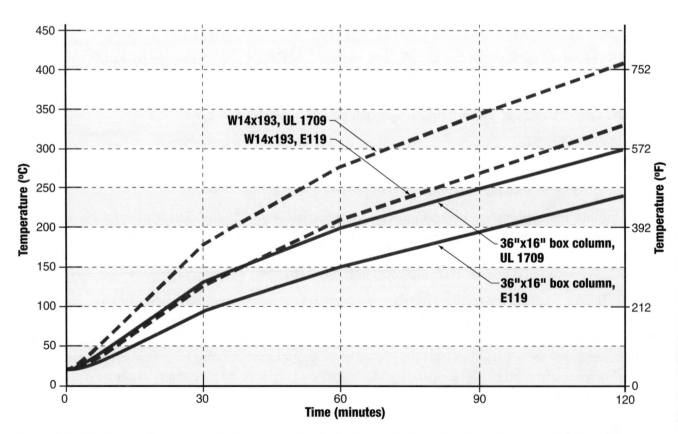

Figure A-10 *Steel temperature rise due to fire exposure for steel column protected with 1 inch of spray-applied fireproofing.*

Fully developed building fires can generally attain average gas temperatures throughout the room containing the fire in excess of 1,000 °C (1,800 °F). The temperature measurements acquired in experiments involving office furnishings conducted by DeCicco, et al. (1972) in the Hudson Terminal Building (30 Church Street, New York), along with the two time-temperature curves from the standard tests is presented in Figure A-3. Temperature development in the first 5 minutes in the room space is notably similar in the experiment with that in ASTM E1529, UL 1709, and the bare thermocouple temperatures for ASTM E119.

Greater temperatures may be acquired locally in a room and especially within flames. Research has indicated that, in the center of flames generated by relatively small fires, temperatures may approach 1,300 °C (2,400 °F) (Baum and McCaffrey 1988). For larger fires, where radiation losses may be reduced, it is conceivable that fire temperatures could reach 1,400 °C (2,550 °F), although this has not been confirmed experimentally.

A.3.1.7 Factors Affecting Performance of Steel Structures in Fire

Several factors influence the behavior of steel structures exposed to fire. The more significant factors are discussed in the following sections.

Loading: One of the major factors that influence the behavior of a structural steel member exposed to fire is the applied load (Fitzgerald 1998, Lie 1992). A loss of structural integrity is expected when the applied loading exceeds or is equal to the ultimate strength of the member. The limiting temperature and the fire resistance of the member increases if the applied load decreases. Traditional fire resistance tests apply a load that results in the maximum allowable stress on the structural member resistance.

Connections: Beam-to-column connections in modern steel-framed buildings may be either of bolted or welded construction, or a combination of these types. Most are designed to transmit shears from the beam to the column, although some connections are designed to provide flexural restraint between the beam and column, as well, in which case they are termed "moment resisting." When moment-resisting connections are not provided in a building, diagonal bracing or shear walls must be provided for lateral stability. When fire-induced sagging deformations occur in simple beam elements with shear connections, the end connections provide restraint against the induced rotations and develop end moments, reducing the mid-span moments in the beams, as well as the tensile catenary action. The moment and tension resisted by connections reduces the effective load ratio to which the beams are subjected, thereby enhancing the fire resistance of the beams as long as the integrity of the connection is preserved. This beneficial effect is more pronounced in large multi-bay steel frames with simple connections. Connections are generally not included as part of the assembly tested in traditional fire resistance tests. Further, most modeling efforts assume that the pre-fire characteristics of a connection are preserved during the fire exposure.

The investigating team observed damaged connections in WTC 5. For example, distorted bolts and bolt holes were found. The performance of connections seem to often determine whether a collapse is localized or leads to progressive collapse. In the standard fire tests of structural members, the member to be tested is wedged into a massive restraining frame. No connections are involved. The issue of connection performance under fire exposure is critical to understanding building performance and should be a subject of further research.

End Restraint: The structural response of a steel member under fire conditions can be significantly enhanced by end restraints (Gewain and Troup 2001). For the same loading and fire conditions, a beam with a rotational restraint at its ends deflects less and survives longer than its simply supported, free-to-expand counterpart. The addition of axial restraint to the end of the beam results in an initial increase in the deflections, due to the lack of axial expansion relief. With further heating, however, the rate of increase in deflection slows.

Effectiveness of Fireproofing: The acceptability of a particular fireproofing material as an insulator is examined as part of ASTM E119. The fireproofing material should form a stable thickness of insulating cover for the steel. Mechanical or impact damage to the fireproofing material prior to the fire exposure that results in a loss of insulating material reduces the ability of the material to act as an insulator (Ryder, et al. 2002). During the fire exposure in the ASTM E119 tests, fireproofing material may fall off as a result of thermal strains caused by differing amounts of expansion in the fireproofing and steel, excess curvature of the steel, or decomposition of the fireproofing material. If the fall-off occurs early in the test or fire exposure, the performance of the assembly is likely to be unsatisfactory. However, if the fireproofing material falls off late in the test or at the time when the fire is declining in intensity, the impact of the lost protection may not be significant. Several test methods other than ASTM E119 can be followed to assess the performance characteristics of fireproofing material. These tests are indicated in Table A.4.

Both the sprayed fiber and, to a lesser extent, cementitious materials, can sometimes fail to adhere to the steel, be mechanically damaged, or otherwise be degraded when exposed to a fire. The current quality control testing of adhesion/cohesion and density, while helpful, does not solve the problem of assuring that the fireproofing will be present at the time of a fire and function throughout the duration of the fire exposure. Other factors that can affect the durability and performance of fireproofing include resistance to abrasion, shock, vibration, and high temperatures.

Sprinklers: Sprinkler systems can be very effective in protecting all structures from the effects of fire. Automatic sprinkler systems are considered to be an effective and economical way to apply water promptly to control or suppress a fire. In the event of fire in a building, the temperature rise in the structural members located in the vicinity of sprinklers is limited. Therefore, the fire resistance of such members is enhanced. The sprinkler piping is sized considering all sprinklers in a design area of operation that are discharging water. For office buildings, typical areas of operation are approximately 1,500 to 2,500 square feet. Should a fire involve an area larger than the area of operation, the water supply may be overwhelmed, thereby negatively impacting the effectiveness of the sprinkler system.

Table A.4 Test Methods for Spray-applied Fireproofing Materials

Standard	Title
ASTM E605	*Thickness and Density of Sprayed Fire-Resistive Materials Applied to Structural Members.*
ASTM E736	*Cohesion/Adhesion of Sprayed Fire-Resistive Materials Applied to Structural Members*
ASTM E759	*Effect of Deflection of Sprayed Fire-Resistive Materials Applied to Structural Members*
ASTM E760	*Effect of Impact on the Bonding of Sprayed Fire-Resistive Materials Applied to Structural Members*
ASTM E761	*Compressive Strength of Sprayed Fire-Resistive Materials Applied to Structural Members*
ASTM E859	*Air Erosion of Sprayed Fire-Resistive Materials Applied to Structural Members*
ASTM E937	*Corrosion of Steel by Sprayed Fire-Resistive Materials Applied to Structural Members*

Structural Interaction: In contrast to an isolated member exposed to fire, the way in which a complete structural building frame performs during a fire is influenced by the interaction of the connected structural members in both the exposed and unexposed portions of the building. This is beneficial to the overall behavior of the complete frame, because the collapse of some of the structural members may not necessarily endanger the structural stability of the overall building. In such cases, the remaining interacting members develop an alternative load path to bridge over the area of collapse. This is a current area of research and is not addressed by traditional fire resistance tests.

Tensile Membrane Action: A tensile membrane (catenary) action can be developed by metal deck and reinforced concrete floor slabs in a steel-framed building whose members are designed and built to act compositely with the concrete slab (Nwosu and Kodur 1999). This action occurs when the applied load on the slab is taken by the steel reinforcement, due to cracking of the entire depth of concrete cross-section or heating of supporting steel members beyond the critical temperature. Tensile membrane action enhances the fire resistance of a complete framed building by providing an alternative load path for structural members that have lost their loadbearing capacity.

Temperature Distribution: Depending on the protective insulation and general arrangements of members in a structure, steel members will be subjected to temperature distributions that vary along the length or over the cross-section. Members subjected to temperature variation across their sections may perform better in fire than those with uniform temperature. This is due to the fact that sections with uniform temperatures will attain their load capacity at the same time. However, in members subjected to non-uniform temperature distribution, a thermally induced curvature will occur to add to the deflections due to applied loads and some parts will attain the load limit before the others. Temperature distributions within structural members may be attained if the member is part of a wall or floor-ceiling assembly where the fire exposure is applied only to one side.

A.3.2 Effect of Fire on Concrete

A.3.2.1 General

Concrete is one of the principal materials widely used in construction and, in fire protection engineering terminology, is generally classified as Group L (loadbearing) building material: materials capable of carrying high stresses. The word concrete covers a large number of different materials, with the single common feature that they are formed by the hydration of cement. Because the hydrated cement paste amounts to only 24 to 43 volume percent of the materials present, the properties of concrete may vary widely with the aggregates used.

Traditionally, the compressive strength of concrete used to be around 20-50 MPa, which is referred to as normal-strength concrete . Depending on the density, concretes are usually subdivided into two major groups: (1) normal-weight concrete, made with normal-weight aggregate, with densities in the 2,200 to 2,400 kg/m^3 range, and (2) lightweight concrete, made with lightweight aggregate, with densities between 1,300 and 1,900 kg/m^3.

The floor slabs at WTC 1 and WTC 2 (as well as in most of the WTC buildings and vicinity) were made of concrete made of metal deck. The floor construction typically consisted of 4 inches of lightweight concrete fill on corrugated metal deck. Hence, the discussion here is focused on lightweight concrete.

A.3.2.2 Properties of Lightweight Concrete

As with steel, concrete loses strength with temperature, though some concretes maintain their ambient temperature strength up to a greater temperature than structural steel. Some lightweight concretes may not exhibit the same level of performance as normal weight concretes under severe fire conditions. In these concretes, spalling under fire conditions is one of the major concerns. The fire resistance of lightweight concrete structural members is dependent on spalling characteristics in addition to thermal and mechanical properties of lightweight concrete at elevated temperatures.

A great deal of information is available in the literature on the properties of lightweight concrete (Abrams 1979, ACI 1989, Lie 1992, Kodur and Harmathy 2002). The modulus of elasticity (E) of various concretes at room temperature may fall within a very wide range, 5.0×10^3 to 50.0×10^3 MPa, dependent mainly on the water-cement ratio in the mixture, the age of concrete, and the amount and nature of the aggregates. The modulus of elasticity decreases rapidly with the rise of temperature, and the fractional decline does not depend significantly on the type of aggregate (Kodur 2000)(see Figure A-11; E_0 in the figure is the modulus of elasticity at room temperature).

The compressive strength (σ_u) of lightweight concrete can vary within a wide range and is influenced by the same factors as the modulus of elasticity. For conventionally produced lightweight concrete (at the time of the WTC construction in 1970s), the strength at room temperature usually was in the 20 to 40 MPa range. The variation of the compressive strength with temperature is presented in Figure A-12 for two lightweight aggregate concretes, one of which is made with the addition of natural sand (Kodur 2000); $(\sigma_u)_0$ in the figures refers to the compressive strengths of concrete at room temperature). The strength decrease is minimal up to about 300 °C (570 °F); above these temperatures, the strength loss is significant.

Generally, lightweight concrete has a lower thermal conductivity, lower specific heat, and lower thermal expansion at elevated temperatures than normal-strength concrete. As an illustration, the usual ranges of variation of the specific heat for normal-weight and lightweight concretes are shown in Figure A-13.

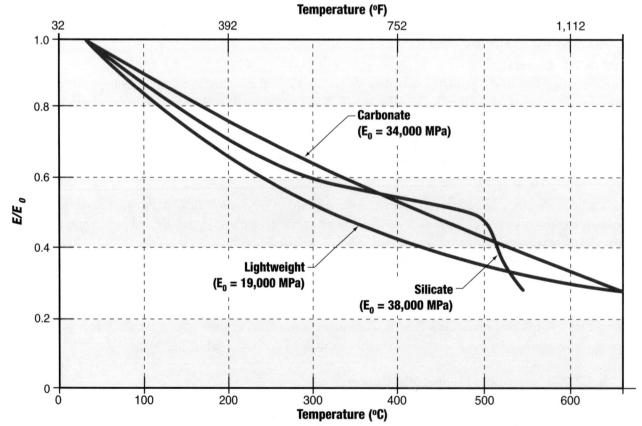

Figure A-11 The effect of temperature on the modulus of elasticity strength of different types of concretes (Kodur and Harmathy 2002).

APPENDIX A: *Overview of Fire Protection in Buildings*

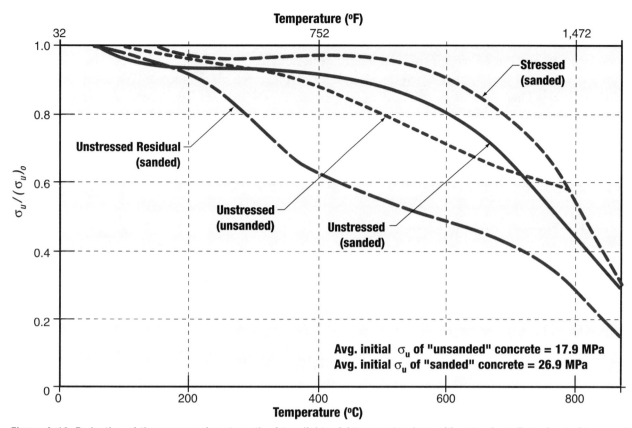

Figure A-12 Reduction of the compressive strength of two lightweight concretes (one with natural sand) at elevated temperatures (Kodur and Harmathy 2002).

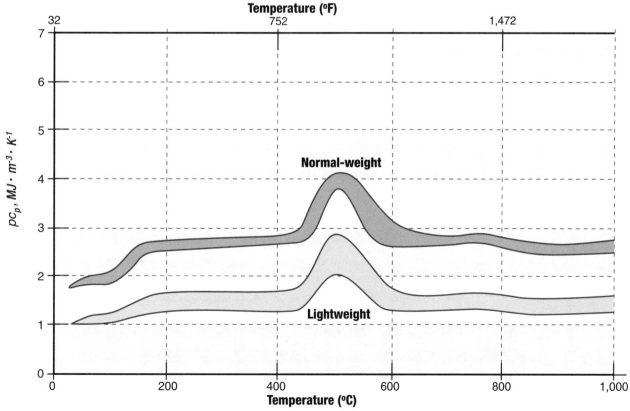

Figure A-13 Usual ranges of variation for the volume-specific heat of normal-weight and lightweight concretes (Kodur and Harmathy 2002).

Spalling is defined as the breaking of layers (pieces) of concrete from the surface of the concrete elements when it is exposed to high and rapidly rising temperatures. The spalling can occur soon after exposure to heat and can be accompanied by violent explosions, or it may happen when concrete has become so weak after heating that, when cracking develops, pieces fall off the surface. The consequences may be limited as long as the extent of the damage is small, but extensive spalling may lead to early loss of stability and integrity due to exposed reinforcement and penetration of partitions.

The extent of spalling is influenced by fire intensity, load intensity, strength and porosity of concrete mix, density, aggregate type, and internal moisture content of the concrete. Significant spalling can occur if the concrete has high moisture content and is exposed to a rapid growth fire.

A.3.3 Fire and Structural Modeling

Fire protection provided in accordance with building codes is based on laboratory tests that have no correlation with actual fires. Through the use of numerical models, the fire protection design of structural members can be determined given the exposure conditions from selected fire scenarios.

Building code requirements for fire resistance design are currently based on the presumed duration of a standard fire as a direct function of fire load, building occupancy, height, and area. The severity of actual fires is determined by additional factors, which are not now considered in current building codes except as an alternate material method or equivalency when accepted by the enforcing official. Recent fire research provides a basis for designing fire protection for structural members by analytical methods and is becoming more acceptable to the building code community. In recent years, the use of numerical methods to calculate the fire resistance of various structural members has begun to gain acceptance. These calculation methods are reliable and cost-effective and can be applied to analyze performance in a specific situation (Milke 1999). The Eurocodes currently describe a calculation method for assessing the performance of steel members exposed to actual fires. There are three analyses that need to be conducted in a numerical assessment of fire resistance:

- model fire development
- model thermal response of assemblies
- model structural response of assemblies

Fire development is modeled to describe the heating exposure provided by the fire. Next, the thermal response analysis consists of predicting the temperature rise of structural members. Finally, an analysis of structural performance can be conducted to determine the structural integrity or load carrying capacity of the fire-exposed structural members. Such an analysis needs to account for thermally-induced deformations and property changes.

The analysis of the WTC buildings and the evaluation of other existing and future tall buildings could involve both fire and structural modeling. Both mathematical and scale modeling, along with validation tests, may be needed. In terms of the numerical modeling, it is currently possible to assemble a model package that reasonably predicts the impact of the fire on strength, elongation, spalling, and other properties related to the structural stability of the buildings involved. Currently, the available models for air movement (to the fire), fire growth and the resulting environmental condition in the space, breaking of windows, heat transfer through materials (e.g., fireproofing), and temperature rise in structural elements operate independently of each other and generally do not share data. In the future, combined fire-structural models may emerge that can interactively feed the output from heat transfer analysis models to structural analysis routines on a time basis as the simulated fire progresses, with return feed to the fire models of any changes (pertinent to the fire model) that the structural computations predict, such as changes in ventilation characteristics. The combined fire-structural model(s) would permit extending the analysis of the impact of this incident to other scenarios, such as fire alone or other combinations of multiple simultaneous impacts (e.g., fire with wind, earthquake) on buildings.

Although the current models are based on sound physics, the state of the art of existing models involves uncertainties. Most of the models needed to supply the structural designer with case-specific data on temperatures of the exposed structural elements in unit area increments matching the finite elements selected for structural analysis exist. However, most of these models are as yet only partially validated.

A.4 Life Safety

The matter of high-rise evacuation has become preeminent in fire and building discussions since September 11, 2001, as a result of the fatalities of over 3,000 building occupants and emergency personnel. Life safety is provided to building occupants by either giving them the opportunity to evacuate or be protected in place. Basic life safety principles include notification, evacuation (including relocation to other floors), and protection in place (SFPE 2000).

Notification: Occupants need to be notified promptly of an emergency. In addition, communication systems should be provided that allow automatic messages to be transmitted to occupants to given them specific instructions on how to respond. These messages may also be delivered over public address systems by building safety managers or fire suppression personnel.

Evacuation: This aspect involves providing people with the means to exit the building. The egress system involves the following considerations:

- Capacity - A sufficient number of exits of adequate width to accommodate the building population need to be provided to allow occupants to evacuate safely.

- Access - Occupants need to be also to access an exit from wherever the fire is, and in sufficient time prior to the onset of untenable conditions. Alternative exits should be remotely located so that all exits are not simultaneously blocked by a single incident.

- Protected Escape Route - Exits need to be protected by fire-rated construction to limit the potential for fire and heat to impact these routes until the last occupant can reach a place of safety. In addition, such routes may also be smoke protected to limit smoke migration into the route.

In general, the means of egress system is designed so that occupants travel from the office space along access paths such as corridors or aisles until they reach the exit. An exit is commonly defined as a protected path of travel to the exit discharge (NFPA 101 2000). The stairways in a high-rise building commonly meet the definition of an exit. In general, the exit is intended to provide a continuous, unobstructed path to the exterior or to another area that is considered safe. Most codes require that exits discharge directly to the outside. Some codes, such as NFPA 101, permit up to half of the exits to discharge within the building, given that certain provisions are met.

Design considerations for high-rise buildings relative to these two options involve several aspects, including design of means of egress, the structure, and active fire protection systems, such as detection and alarm, suppression, and smoke management.

There is no universally accepted standard on emergency evacuation. Many local jurisdictions through their fire department public education programs have developed comprehensive and successful evacuation planning models, but unless locally adopted, there is no legal mandate to exercise the plans. Among the cities that have developed comprehensive programs are Seattle, Phoenix, Houston, and Portland, Oregon.

Protect in Place: The protect in place strategy is commonly employed in high-rise buildings. Occupants either remain in an area enclosed in fire rated construction or move to such a location. This approach is especially important for mobility impaired individuals. Building construction and fire protection systems are employed to protect occupants from fire and smoke spread for the duration of the incident or until rescued.

In some cases, occupants may be moved from one location to a location of relative safety while they await rescue. The Americans with Disabilities Act (ADA) of 1990 (42 USC 12181), in its design guidelines for new construction since 1993, requires that each floor in a building without a supervised sprinkler system must contain an "area of rescue assistance" (i.e., an area with direct access to an exit stairway where people unable to use stairs may await assistance during an emergency evacuation). In existing buildings, the ADA makes no reference to occupant evacuation other than to prohibit unnecessary physical barriers to mobility.

Additional information about courses and publications on emergency evacuation can be obtained at http://www.usfa.fema.gov.

A.4.1 Evacuation Process

Two methods are followed for the evacuation of buildings. One method consists of evacuating all occupants simultaneously. Alternatively, occupants may be evacuated in phases, where the floor levels closest to the fire are evacuated first, then other floor levels are evacuated on an as needed basis. Phased evacuation is instituted to permit people on the floor levels closest to the fire (i.e., those with the greatest hazard) to enter the stairway unobstructed by queues formed by people from all other floors also being in the stairway. Those who are below the emergency usually are encouraged to stay in place until the endangered people from above are already below this respective floor level. Generally, phased evacuation is followed in tall buildings, such as WTC 1 and WTC 2.

A.4.2 Analysis

A fairly simplistic model can be applied to develop a first order approximation of the time required to evacuate a high-rise building. The model is described by Nelson and MacLennan in the SFPE Handbook of Fire Protection Engineering. The following calculations are based on several major assumptions:

- All persons start to evacuate at the same time and hence no pre-movement time is considered (e. g., talking to coworkers, turning off computers, putting on coats).

- Occupant travel is not interrupted to make decisions or communicate with other individuals involved.

- The persons involved are free of any disabilities that would significantly impede their ability to keep up with the movement of the group. This includes any temporary disabilities as a result of fatigue.

- Firefighters coming into the stairway do not impose a significant impact on the flow rate of occupants traveling down the stairs.

- The controlling feature of the flow rate of people from the building is the door at the bottom of the exit stairway. This assumes that people develop a queue in the stairway that ends at the doorway at the base of the stairway. Also, the time for the first people to form the queue is assumed to be much less than the total evacuation time.

- The density of the people traveling through the doorway is in the range of observed values (i.e., 6-10 ft^2/person). As such, the flow rate per foot of effective width for each doorway would be anticipated to be in the range of 18 to 24 persons/min (see Figure A-14). Consequently, the flow rate from each doorway in the World Trade Center buildings would have been on the order of 30 to 50 persons/min.

APPENDIX A: *Overview of Fire Protection in Buildings*

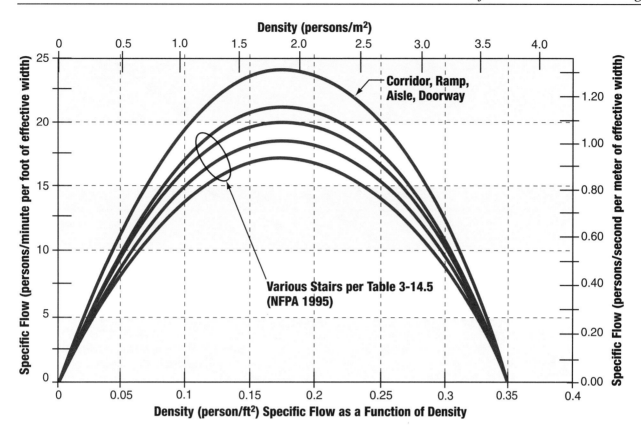

Figure A-14 Specific flow rate as a function of density (SFPE Handbook of Fire Protection Engineering).

Given these assumptions, the results presented in Figure A-15 relate to a lower limit of the time expected to evacuate the WTC towers. There were three exit stairways serving most floors of the WTC towers. Below the impact area, all stairways appeared to be available. The number of people in each building on the morning of September 11, 2001, is not known. Therefore, a range of occupant loads is included in Figure A-15.

By all indications, it was instantly apparent to the building occupants that evacuation was necessary, so very little time was likely to have transpired in pre-movement activities. The time for the leading edge of the evacuees to reach the stairs and to descend from the lowest occupied floor (7) to the discharge doors on floors 1 and 2 is estimated to have taken about 3 minutes until the steady human flow reached its capacity. The sense of urgency in the evacuees is estimated to have maintained the egress flow at or near the theoretical maximum for stair exit flow (i.e., 24 persons/minute per foot).

The two end stairs were 44 inches wide and the center stair was 56 inches wide. Each stair had a single 36-inch-wide exit door at its discharge level. As such, the effective width for each stair door was 24 inches (2 feet). The expected steady flow rate from the stair doorways was 48 persons/minute. Based on an available egress time of 90 minutes in WTC 1 and 50 minutes in WTC 2, the number of persons who could have exited through the stairs is estimated to be up to 13,000 for WTC 1 and up to 7,200 for WTC 2. These estimates do not include any persons who used elevators, were on the 2nd (Plaza) level or lower in the buildings at the time, or initiated evacuation in WTC 2 immediately after the impact of WTC 1.

APPENDIX A: *Overview of Fire Protection in Buildings*

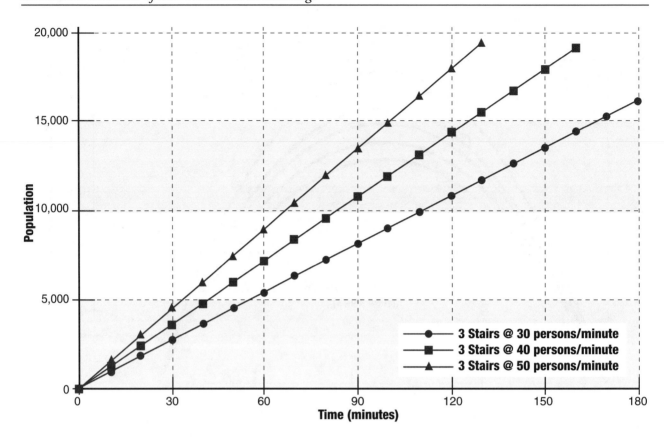

Figure A-15 Estimated evacuation times for high-rise buildings.

A.5 References

Abrams, M. 1979. "Behavior of Inorganic Materials in Fire," *Design of Buildings for Fire Safety*, ASTM 685. Philadelphia.

ACI. 1989. *Guide for Determining the Fire Endurance of Concrete Elements*, ACI-216-89. American Concrete Institute, Detroit.

ASCE/SFPE. 1999. *Standard Calculation Methods for Structural Fire Protection*, ASCE/SFPE 29. American Society of Civil Engineers, Washington, DC, and Society of Fire Protection Engineers, Bethesda, MD.

ASTM. 2001a. *Standard Terminology of Fire Standards*, ASTM E176. American Society for Testing and Materials, West Conshohocken, PA.

ASTM. 2001b. *Standard Test Methods for Determining Effects of Large Hydrocarbon Pool Fires on Structural Members and Assemblies*, ASTM E1529. American Society for Testing and Materials, West Conshohocken, PA.

ASTM. 2000. *Standard Test Methods for Fire Tests of Building Construction and Materials*, ASTM E119, American Society for Testing and Materials, West Conshohocken, PA.

Baum, H. R., and McCaffrey, B. J. 1988. "Fire Induced Flow Field: Theory and Experiment." *Proceedings of the 2nd International Symposium, International Association for Fire Safety Science*. 129-148.

BSI. 2000. *Structural use of steelwork in building. Code of practice for design. Rolled and welded sections*, BS 5950-1:2000. British Standards Institute, London.

DeCicco, P.R., Cresci, R.J., and Correale, W.H., 1972. *Fire Tests, Analysis and Evaluation of Stair Pressurization and Exhaust in High Rise Office Buildings*. Brooklyn Polytechnic Institute.

Fitgerald, R.W. 1998. "Structural Integrity During Fire," *Fire Protection Handbook*, 18th Edition. A.E. Cote, National Fire Protection Association, Quincy, MA.

Gerlich, J.T. August 1995. *Design of Loadbearing Light Steel Frame Walls for Fire Resistance*, Fire Engineering Research Report 95/3. University of Canterbury, Christchurch, NZ.

Gewain, R.G., and Troup, E.W.J. 2001. "Restrained Fire Resistance Ratings in Structural Steel Buildings," *AISC Engineering Journal*.

Ingberg, S. H. 1928. "Tests of the Severity of Fires," *Quarterly NFPA*. 22, 3-61.

Kodur, V.K.R., and Harmathy, T.Z. 2002. "Properties of Building Materials," *SFPE Handbook of Fire Protection Engineering*, 3rd edition. P.J. DiNenno, National Fire Protection Association, Quincy, MA.

Kodur, V.K.R., and Lie, T.T. 1995. "Fire Performance of Concrete-Filled Steel Columns," *Journal of Fire Protection Engineering*, Vol. 7.

Kodur, V.K.R. 2000. "Spalling in high strength concrete exposed to fire - concerns, causes, critical parameters and cures," *Proceedings: ASCE Structures Congress*. Philadelphia, U.S.A.

Lie, T.T. 1992. *Structural Fire Protection: Manual of Practice*, No. 78, ASCE, New York.

Madrzykowski, D. 1996. "Office Work Station Heat Release Rate Study: Full Scale vs. Bench Scale," Interflam '96. *Proceedings of the 7th International Interflam Conference*. Cambridge, England.

Magnusson, S. E., and Thelandersson, S. 1970. "Temperature-Time Curves of Complete Process of Fire Development in Enclosed Spaces," *Acts Polytechnica Scandinavia*.

Makelainen, P., and Miller, K. 1983. *Mechanical Properties of Cold-Formed Galvanized Sheet Steel Z32 at Elevated Temperatures*. Helsinki University of Technology, Finland.

Milke, J.A. 1995. "Analytical Methods for Determining Fire Resistance of Steel Members," *SFPE Handbook of Fire Protection Engineering*, 2nd edition. P.J. DiNenno, National Fire Protection Association, Quincy, MA.

Milke, J.A. 1999. "Estimating the Fire Performance of Steel Structural Members," *Proceedings of the 1999 Structures Congress*, ASCE. 381-384.

NFPA 101. 2000. "Life Safety Code." National Fire Protection Association, Quincy, MA.

Nwosu, D.I., and Kodur, V.K.R.1999. "Behaviour of Steel Frames Under Fire Conditions," *Canadian Journal of Civil Engineering*. 26, 156-167.

Ryder, N.L., Wolin, S.D., and Milke, J.A. 2002. "An Investigation of the Reduction in Fire Resistance of Steel Columns Caused by Loss of Spray Applied Fire Protection," *Journal of Fire Protection Engineering*. To be published.

SFPE. 2000. *SFPE Engineering Guide to Performance Based Fire Protection Analysis and Design of Buildings*. March.

SFPE. 1995a. *The SFPE Handbook of Fire Protection Engineering*, 2nd Edition. p. A-43 - A-44, Table C-4. NFPA, Quincy, MA.

SFPE. 1995b. *The SFPE Handbook of Fire Protection Engineering*, 2nd Edition. p. 3-78 - 3-79, Table 3-4.11. NFPA, Quincy, MA.

Thomas, I.R., and Bennetts, I. 1999. "Fires in Enclosures with Single Ventilation Openings: Comparison of Long and Wide Enclosures." *Proceedings of the 6th International Symposium on Fire Safety Science*. University of Portiers, France.

Wang, Y.C., and Kodur, V.K.R. 2000. "Research Towards the Use of Unprotected Steel Structures," *ASCE Journal of Structural Engineering*. Vol. 126, December.

John Fisher
Nestor Iwankiw

B *Structural Steel and Steel Connections*

B.1 Structural Steel

This appendix focuses mainly on the structural steel and connections in the WTC towers (WTC 1 and WTC 2), but column-tree connections in WTC 5 are also considered. Other WTC structures were fabricated from ASTM A36 and A572 grade steels, and their structural framing and connections are discussed in prior chapters.

The structural steel used in the exterior 14-inch by 14-inch columns that were spaced at 3 feet 4 inches on center around the entire periphery of each of the WTC towers was fabricated from various grades of high-strength steel with minimum specified yield stress between 36 kips per square inch (ksi) and 100 ksi (PATH-NYNJ 1976). Column plate thickness varied from 1/4 inch to 5/8 inch in the impact zone of WTC 1 for floors 89-101, and from 1/4 inch to 13/16 inch in the impact zone of WTC 2 for floors 77-87. Spandrel beams at each floor level were fabricated of matching steel and integrated into the columns as the columns and spandrel sections were prefabricated into trees. These trees were three columns wide and one to three stories high. The cross-sectional shape of the columns can be seen in Figure B-1. These varied in length from 12 feet 6 inches to 38 feet, depending on the plate thickness and location.

Figure B-1 Exterior column end plates.

The three columns in a panel were generally fabricated from the same grade of steel. The yield stress varied from 50 ksi to 100 ksi in increments of 5 ksi up to 90 ksi. Although most of the time the same grade of steel was used in all three columns, sometimes a column was fabricated from different grades. The difference was up to 15 ksi (i.e., 75 ksi, 85 ksi, and 90 ksi). The core columns were box sections fabricated from A36 steel plate and were 36 inches x 14–16 inches with plate thickness from 3/4 inch to 4 inches. Above floor 84, rolled or welded built-up I-shaped sections were used.

The floor system was supported by 29-inch-deep open-web joist trusses with A36 steel chord angles and steel rod diagonals. Composite 1-1/2-inch, 22-gauge metal floor deck ran parallel to double trusses that were spaced at 6 feet 8 inches. The floor deck was also supported by alternate intermediate support angles and transverse bridging trusses that were spaced at 3 feet 4 inches. The bridging truss also framed into some periphery columns. Figure 2-2 (in Chapter 2) shows the layout of a typical floor. Because 13-foot-wide and 20-foot-wide modular floor units were prefabricated for construction, the outside two trusses shared a common top chord seat connection with adjacent panels. All double trusses were attached to every other periphery column by a seat angle connection and a gusset plate that was welded to the spandrel and top chord. Therefore, all truss supports had two trusses attached to the seat connection. A single bolt was used for each truss sharing a seat connection. The bottom chord of each pair of trusses was attached to the spandrel with visco-elastic dampers that had a slip capacity of 5 kips. At the core, the trusses were connected to girders that were attached to the box or H-shaped core columns by beam seats welded to the column faces.

B.2 Mechanical Properties

Nearly all of the steel plate was produced in Japan to ASTM standards or their equivalent. None of the mill test reports were available that describe the mechanical properties and chemical composition of the steel used in the WTC structures. Approximately 100 potentially helpful steel pieces were identified at the four salvage yards that had contracts to obtain and process the WTC steel debris. These pieces have been removed and transported to the National Institute of Standards and Technology (NIST) in Gaithersburg, Maryland, for storage and further study. No coupons were taken or tested to check material conformance with specification of any plate, rolled section, bolt, weld, reinforcing steel, or concrete. Visual examination of the debris did not identify any apparent deficiencies in the structural materials and connectors.

In lieu of actual WTC steel properties, typical stress-strain curves characteristic of 3 of the 12 steels used in the design and construction of the WTC complex are shown in Figure B-2 for three ASTM-designation steels with minimum specified yield strengths of 36 ksi (A36), 50 ksi (A441), and 100 ksi (A514). In general, as the yield strength of the steel increases, the yield-to-tensile-strength ratio (Y/T) also increases. For A36 steel, Y/T is approximately 0.6, whereas for A514 steel, Y/T is approximately 0.9. The yield plateau for five steels (yield points 36, 50, 65, 80, and 100 ksi) can be highly variable for structural steels, as is apparent from a comparison of the expanded initial portions of the five steels shown in Figure B-3. At the higher yield strength associated with quenched and tempered alloy steels, there may not be a distinct yield plateau; instead, the steels exhibit gradual yielding and nonlinear behavior with strain hardening.

High strain rates tend to increase the observed yield strength and tensile strength of steel, but may also reduce the ductility. There is a greater influence on the yield point than on the tensile strength. Figure B-4 compares the effect of a very high strain rate (100 in/in/sec) for a mild carbon steel with a more usual test speed of 850 micro in/in/sec. In this example, the yield point more than doubled, whereas the tensile strength was increased about 27 percent, and the Y/T ratio approached unity.

In fracture toughness tests where rapid load toughness is determined, the dynamic yield strength of certain steels can be estimated by the following equation taken from ASTM E1820 (ASTM 1999):

$$\sigma_{yd} = \sigma_{ys} + \frac{174{,}000}{(T+460)\log_{10}(2 \times 10^7\, t)} - 27.2\ ksi \tag{B-1}$$

APPENDIX B: *Structural Steel and Steel Connections*

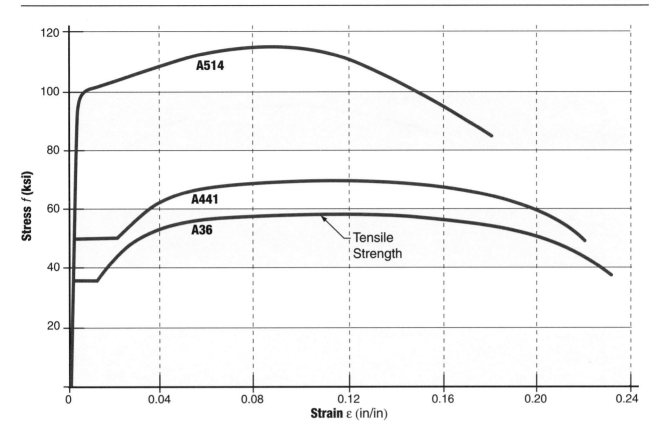

Figure B-2   Tensile stress-strain curves for three ASTM-designation steels (Brockenbrough and Johnston 1968, Tall 1974).

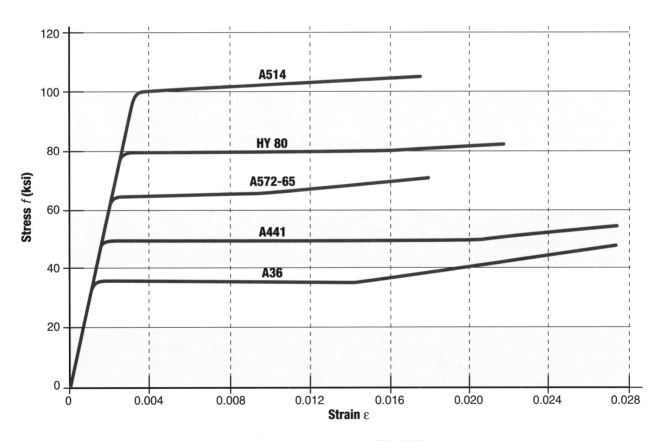

Figure B-3   Expanded yield portion of the tensile stress-strain curves (Tall 1974).

APPENDIX B: *Structural Steel and Steel Connections*

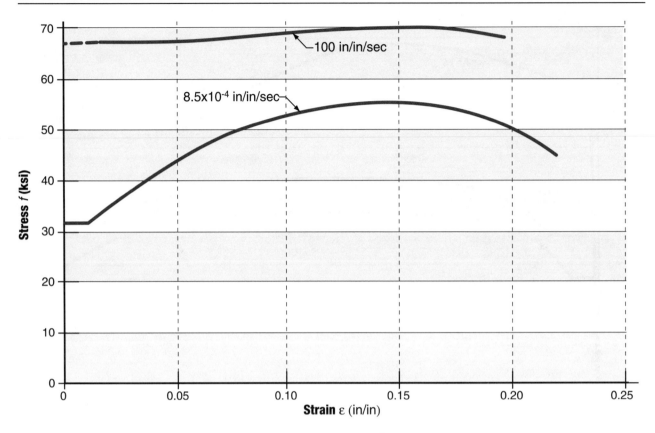

Figure B-4   Effect of high strain rate on shape of stress-strain diagram (Tall 1974).

Where σ_{ys} is room temperature static yield strength in ksi, t = loading time in milliseconds, and T is the test temperature in °F.

In *Making, Shaping, and Treating of Steel* (USX 1998), it is noted that a tenfold increase in rate of loading increased a 0.12 percent carbon steel yield strength by 7 ksi, but the influence on tensile strength was negligible.

High impacts that create notches can also lead to brittle fracture at stresses that are less than the dynamic yield strength. This is also true if triaxial stress conditions exist from constraint.

The high-temperature characteristics of structural steel are discussed in Appendix A.

B.3   WTC 1 and WTC 2 Connection Capacity

B.3.1   Background

Connections are typically designed to transfer the joint forces to which they are subjected. Generally, simple equilibrium models are used to proportion the mechanical or welded connectors and the plate or beam elements used in the connection for the required design loads (Fisher, et al. 1978; Kulak, et al. 1987; Lesik and Kennedy 1990; Salmon and Johnson 1996).

According to available information, steel connections in the WTC structures were designed in accordance with the AISC specifications that were applicable at the time to resist the required design loads. This section focuses on the ultimate limit strengths of the connectors and the various connections that were used to construct the WTC towers. Standard practice is that the design of connectors and connections provide a factor of safety of at least two against the various design strength limit states. Significant deformations can be expected when these limit states are reached.

B.3.2 Observations

1. The exterior tree columns were spliced using bolted end plate connections.

2. All column end plate bolted connections appeared to fail from the unanticipated out-of-plane bending of the column tree sections due to either the aircraft impacts or the deformation and buckling of the unbraced columns as the floor system diaphragms were destroyed by the impacts and fires. The bolts were observed to exhibit classical tensile fracture in the threaded area. Most bolts were also bent in the shank. Figure B-1 shows the column end plates and holes with some fractured and bent bolts. No evidence of plastic deformation was observed in the end plates.

3. Column splice requirements in the AISC Specifications (1963) indicated in Section 1.15.8 that "Where compression members bear on bearing plates and where tier-building columns are finished to bear, there shall be sufficient rivets, bolts, or welding to hold all parts securely in place."

B.3.3 Connectors

The connectors generally used for steel structures are either high-strength bolts or welds. The project specifications indicated that bolts were to meet the ASTM A325 or A490 standards.

Bolts are designed based on their nominal shank area A_b for tension, shear, or some combination. For tension, the nominal strength (per unit of area) of a single bolt is provided by

$$F_n = C_t F_u \tag{B-2}$$

where F_u is the minimum specified tensile strength and $C_t = 0.75$, which is the ratio of the stress area to the nominal shank area. An analysis of A325 bolts produced in the 1960s and 1970s indicated that, on average, the bolts exceeded the minimum specified tensile strength by 18 percent (Kulak, et al. 1987).

For shear, the nominal strength of a single bolt is provided by

$$F_v = C_s F_u \tag{B-3}$$

The average shear coefficient $C_s = 0.62$ for a single bolt. This coefficient is reduced to $C_s = 0.5$ to account for connection lengths up to 50 inches parallel to the line of force. When threads are not excluded from the shear plane, the coefficient C_s is further reduced to $C_s = 0.4$.

Other failure modes are possible as bolts transfer forces from one component into another by bearing and shear of the fastener. This can result in bearing deformations and net section fracture of the connected elements. Other failure modes are shear rupture or bearing strength as bolts shear connected material between the bolt and a plate edge, or block shear, which combines the tension and shear resistance of the connected elements. An example of one of these failure modes can be seen in Figure B-5, which shows a spandrel beam bolted shear connection that has failed in end zone shear as the connection was subjected to moments and/or tensile loading. This indicates that all elements in this example were at their ultimate load capacity when the spandrel connection failed.

The commonly used weld connectors are either fillet welds or groove welds. Complete joint penetration groove welds are designed for the same basic capacity as the connected base metal and match its capacity. Fillet welds and partial joint penetration groove welds are designed to resist a calculated or specified load by sizing for the weld throat area, which is the effective cross-sectional area of the weld.

APPENDIX B: *Structural Steel and Steel Connections*

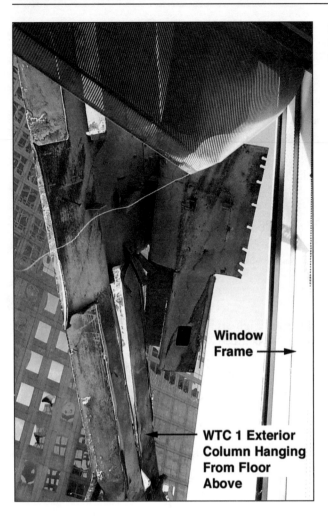

Figure B-5
Column tree showing bolt bearing shear failures of spandrel connection.

The nominal strength of a linear weld group loaded in-plane through the center of gravity is (Fisher, et al. 1978; Lesik and Kennedy, 1990).

$$F_w = 0.6\, F_{EXX}\, (1.0 + 0.5\, \sin^{1.5} \theta) \tag{B-4}$$

where F_{EXX} is the electrode classification number (minimum specified tensile strength) and θ is the angle of loading measured from the longitudinal axis in degrees. Hence, when the load is parallel to the weld, the capacity is $0.6\, F_{EXX}$, and when it is perpendicular to the longitudinal weld axis, it increases to $0.9\, F_{EXX}$ or more. This increased strength of fillet welds transverse to the axis of loading was not recognized in the AISC Specification when the WTC was designed and built.

Figure B-6 shows the shear failure of the fillet welds that connected a built-up wide-flange column to the top end of a box core column. The basic limit states for bolts and welds described by Equations B-2, B-3, and B-4 are used in Chapter J of the AISC Load and Resistance Factor Design Specification along with resistance factors to design structural steel building connections (Fisher, et al. 1978). Additional connection strength design provisions are covered in Chapter K of the same specification for flanges and webs subjected to concentrated forces (Fisher, et al. 1978). Those relationships can also be used to assess the ultimate capacity, or strength, of structural members and connections subjected to concentrated tension and compression forces.

Discontinuities such as porosity and slag seldom cause a significant loss of static strength. Common imperfections are permitted within limits and accommodated by the provisions of the design standards. On the other hand, lack of fusion or cracks can have a major impact on strength and can result in joint

APPENDIX B: *Structural Steel and Steel Connections*

Figure B-6
Shear fracture failure of fillet welds connecting a W-shape column to a box core column.

failure at loads below the design load. This will depend on the size of the discontinuity or defect and its orientation to the applied loads.

B.4 Examples of WTC 1 and WTC 2 Connection Capacity

B.4.1 Bolted Column End Plates

Collapse of the WTC towers resulted in failure of many of the bolts in bolted end plate connections as the columns were subjected to large and unanticipated out-of-plane bending. In the majority of cases, the A325 high-strength bolts reached their tensile capacity and failed in the threaded stress area. The example shown in Figure B-7 examines the flexural capacity of the bolted end plate in a column in the impact area where the column plate thickness was 1/4 inch.

The simple moment capacity of the bolt group is 20 to 30 percent of the plastic moment capacity of a column fabricated from steels with a 50 to 100 ksi yield point, assuming no axial load in the columns. The end plates at the columns splice have a 11-3/4-inch x 14-inch cross-section. The columns are subjected to axial load from the dead load acting on the structure. For the as-built structure, the moments acting on the bolted splice are small, because the splices were located at the column inflection points and the resultant of the applied axial load and moment is within the middle third of the 12-inch-deep bearing connection. Assuming an axial stress of 20 ksi in the column, the corresponding axial force acting on the base plate is 280 kips. As the columns lose lateral support and deform out-of-plane from overloading eccentricities and from the thermal effects, the bending moment acting on the column splice does not introduce significant forces into the bolted end plate connection until the eccentricity exceeds 2 inches. As the eccentricity increases, the applied bending moment will exceed the bolt preload stress when the eccentricity reaches approximately 4 inches. Continued deformation will exceed the ultimate moment capacity of the connection and result in instability as the eccentricity approaches 4.5 inches.

It also should be noted that the column splices were staggered midheight at each floor, as was illustrated in Chapter 2. As a result, two-thirds of the perimeter columns were continuous at each floor's

APPENDIX B: *Structural Steel and Steel Connections*

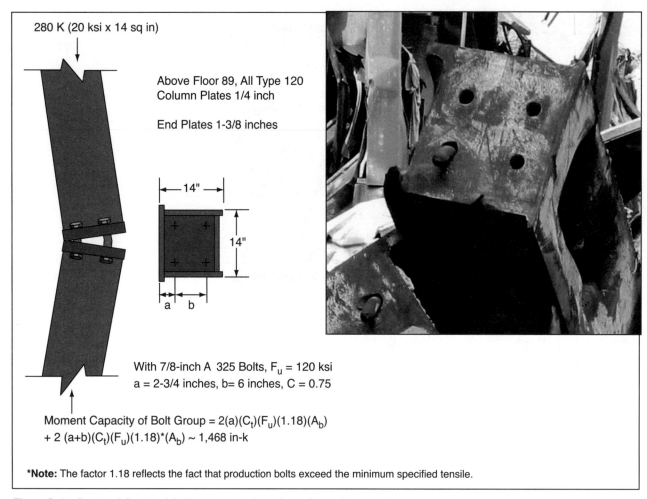

Figure B-7 Bent and fractured bolts at an exterior column four-bolt connection.

midheight elevation. This resulted in staggered failure patterns, as the bolted end plate connections and spandrel beam connections failed during the resulting instability and collapse. The exception to this staggered pattern was the splices at mechanical floors, which were not staggered, and the bolts were supplemented with welds.

B.4.2 Bolted Spandrel Connections

Collapse of the WTC towers resulted in failure of the bolted shear splices that connected the spandrel beams together at each of the prefabricated column trees. Several modes of failure were observed in these connections. Figure B-5 showed an example of bearing strength failure of the spandrel plate. The loading appeared to be a combination of unanticipated moment and tensile loading. The following example examines the shear rupture capacity of one of the bolts. The spandrel plate thickness was assumed to be 3/8 inch, which was observed at the columns with 1/4-inch plate used in Figure B-7.

The ultimate bearing strength capacity is given by (AISC 2001; Fisher, et al. 1978)

$$R_u = L_c t F_u \quad 3.0 \quad dt F_u \quad F_v A_s \tag{B-5}$$

where d is the nominal bolt diameter (7/8 inch), L_c is the clear distance, in the direction of force, between the edge of the hole and the edge of the spandrel plate (1-5/16 inches), F_u is the tensile strength of the spandrel plate, and t is the thickness of the spandrel plate (0.375 inch).

This results in the following bearing capacity of a single bolt

$$R_u = L_c t F_u = 1.3125 \times 0.375 \times 90 = 44.3 \ kips \tag{B-6}$$

This is well below the single shear capacity of the bolt, which is

$$F_v A_s = 0.62 \times 120 \times 1.18 \times 0.6013 = 52.8 \ kips \tag{B-7}$$

Hence, the failure mode observed in Figure B-5 is consistent with the predicted capacity.

B.4.3 Floor Truss Seated End Connection at Spandrel Beam and Core

The floor system supported by 29-inch-deep prefabricated steel trusses consisted of 4 inches of lightweight concrete fill on a 1-1/2-inch corrugated deck that ran parallel with the truss (PATH-NYNJ 1976). As noted in the introduction, alternate truss supports had two joists attached to the seat connection.

Figure B-8 shows the end of the top chords that were connected to every other exterior column/ spandrel beam and the core support channel beams. The top chords were supported on bearing seats at each end of the two trusses. At the exterior column/spandrel beam, a gusset plate was groove-welded to the spandrel face and fillet-welded to the top chord angles. At the bearing seat, two 5/8-inch A325 bolts in 3/4-inch x 1-1/4-inch slotted holes connected the trusses' top chords to the bearing seat with a single bolt in the exterior angle of each truss. The lower chord was attached to the exterior column/spandrel beam with a visco-elastic damping unit connected to a small seat with two 1-inch A490 bolts that provided a slip-resistant connection. The damping unit had a capacity of about 5 kips.

At the core, the top chords were supported by bearing seats with two vertical stiffeners. Two 5/8-inch A325 bolts were installed in 3/4-inch x 1-3/4-inch slotted holes in the seat plate and standard holes in the top chord outside angles.

Figure B-9 shows several of the failure modes of the truss connections to the chord bearing seat and spandrel beam. The gusset plate welded to the spandrel beam and the top chord failed by tensile fracture of the plate. The gusset plate connection was primarily resisting the floor diaphragm support to the column. After fracture, the slotted holes in the seat would allow rigid body motion of the trusses until the 5/8-inch bolts came into bearing. That resulted in partial fracture in the seat of the fillet welds attaching the fill plate to the spandrel beam. The seat angle welded connection to the fill plate remained intact as this separation occurred and final block shear failure developed in the outstanding angle leg at the two slotted bolt holes.

The capacity of the 3/8-inch x 4-inch A36 steel gusset plate can be estimated as:

$$R_u = A_g F_u = (1.5 \ inches^2)(60 \ ksi) = 90 \ kips \tag{B-8}$$

The bearing capacity of the two 5/8-inch bolts connecting the top chord angles to the seat angle is

$$R_u = 2 L_c t F_u = 2(1 \ inch)(0.375 \ inch)(60 \ ksi) = 45 \ kips \tag{B-9}$$

The shear capacity of the two 5/8-inch A325 bolts is

$$2 F_v A_x = 2 \times 0.62 \times 120 \times 1.18 \times 0.307 = 53.9 \ kips \tag{B-10}$$

The block shear rupture strength provided in Chapter J of the AISC LRFD specification (Fisher, et al. 1978) can be used to assess the tensile force that separated the floor joist from the bearing seat

$$R_u = 0.6 F_u A_{nv} + F_u A_{nt} \tag{B-11}$$

APPENDIX B: *Structural Steel and Steel Connections*

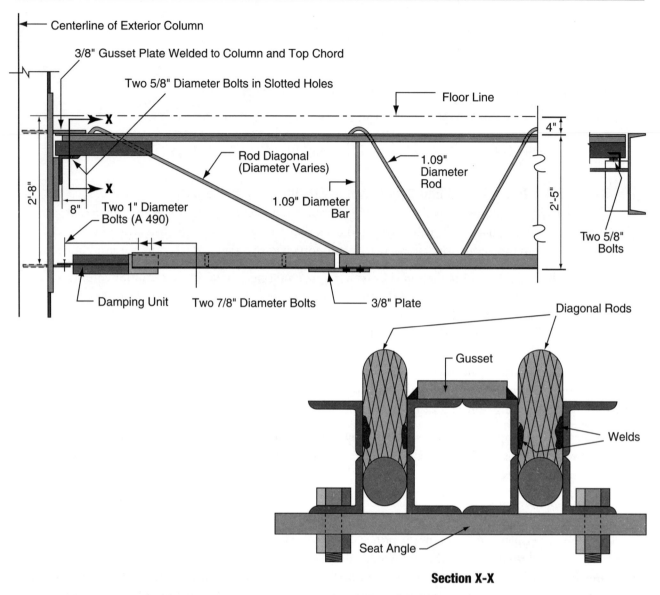

Figure B-8 Typical truss top chord connections to column/spandrel beam and to the core beam.

where A_{nv} is the net shear area = 0.375 inch x 1.5 inches = 0.5632 inch².

The net tension area $A_{nt} = A_{nv}$. Hence, assuming an average tensile strength for A36 steel of Fu = 60 ksi results in a maximum resisting force

$$R_u = 2[0.6\,(60)(0.563) + 60\,(0.563)] = 108 \; kips \tag{B-12}$$

It is probable that, once the 3/8-inch gusset plate fractures, the next lower bound resistance is provided by the bearing capacity of the two 5/8-inch bolts on the beam seat angle. This failure also tore off the ends of the angle even though the tensile capacity of those segments was predicted to be higher.

APPENDIX B: *Structural Steel and Steel Connections*

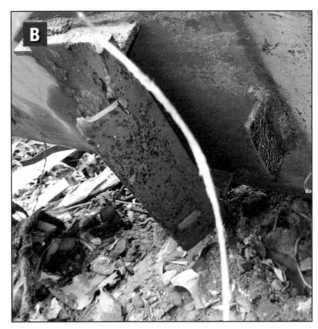

Figure B-9 (A) Visco-elastic damper angles bolted to angle welded to spandrel plate and (B) failed bearing seat connection.

It should also be noted that each truss top chord provided a horizontal diagonal plate brace (1-1/2 inches x 1/2 inch) to the two adjacent columns. These members were welded to welded bracket plates on each adjacent column/spandrel member, as illustrated in Figure B-10. In this case, it would appear that the diagonal plate braces fractured on their gross section or tore the bracket plate. The component of ultimate strength of the two diagonal plate braces normal to the column/spandrel member is about 85 percent of the tensile capacity of braces, which would be 76 kips.

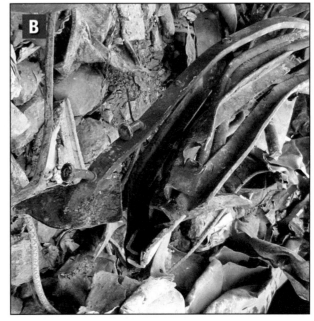

Figure B-10 (A) Bracket plate welded to the column/spandrel plate and (B) horizontal plate brace with shear connectors welded to the failed bracket.

APPENDIX B: Structural Steel and Steel Connections

Many of the bearing seat brackets and the damper angle connections on the column/spandrel beam plate were completely sheared off. Only the weld segments remained on face of the column/spandrel beam plate (Figure B-11). This mode of failure appears to be due to excessive vertical overloads on the floor system. This is in contrast with the failure mode exhibited in Figure B-9 where the bearing seat bracket has pulled away from the column/spandrel plate, after fracture of the top chord gusset plate.

B.4.4 WTC 5 Column-tree Shear Connections

Chapter 4, Section 4.3.2, noted that limited structural collapse had occurred in WTC 5 as a result of failure at the shear connections between the infill beams and the column tree beam stub cantilevers. It is visually apparent from Figures 4-18, 4-19, 4-20, and 4-21 that the fire-weakened structural members formed diagonal tension field failure mechanism in the cantilever beam webs and plastic hinge moments in the cantilever beam near the column face. The following analysis examines the capacity of the cantilever beams and the shear connections between the cantilever and the infill beams.

The magnitude of the shear force acting at the end of the cantilever section can be estimated from the plastic moment capacity and the plastic shear yield capacity (AISC 2001). The plastic moment capacity of a W24x61 steel section is

$$M_p = F_y Z_x \tag{B-13}$$

and the shear yield capacity is

$$V_p = 0.6 F_y d t_w \tag{B-14}$$

where F_y = 40 ksi is the approximate yield strength of the A36 steel sections at room temperature, Z_x is the plastic section modulus, d is the beam depth, and t_w is the web thickness.

Figure B-11
Shear failure of floor truss connections from column/spandrel plate.

The vertical shear capacity of the bolted double shear splice connecting the W18x50 section to the W24x61 cantilever can be estimated for the three-bolt connection from the bearing strength relationship given in Section B.4.2. This gives

$$V_u = 3 R_n = 3(3.0 \, d_b \, t_w \, F_u) \tag{B-15}$$

where d_b = 3/4 inch is the bolt diameter, t_w = 3/8 inch is the web thickness for a W18x50, and F_u = 60 ksi is the approximate tensile strength of A36 steel.

Appendix A indicates the yield strength is 0.9 F_y at 200 °C (392 °F) and the yield strength is 0.5 F_y at 550 °C (1,022 °F). The large plastic deformation observed in the cantilever beam segments suggests that a significant loss of strength developed due to the fire. The fire temperatures reached in WTC 5 are not known, but if it is assumed for the purposes of this analysis that 550 °C (1,022 °F) was reached, this analysis estimates the cantilever plastic moment capacity as

Room Temperature: $\quad M_p = 1.0 \, F_y \, Z_x = 1.0 \, (40)(152) = 6{,}080$ in-k $\tag{B-16a}$

550 °C (1,022 °F): $\quad M_p = 0.5 \, F_y \, Z_x = 0.5 \, (40)(152) = 3{,}040$ in-k $\tag{B-16b}$

This corresponds to a shear force of about 126 kips at room temperature acting at the end of the 4-foot cantilever, which would be reduced to 63 kips at 550 °C (1,022 °F). The limiting shear yield strength of the cantilever is

Room Temperature: $\quad V_p = 0.6 \, (1.0 \times 40)(23)(0.419) = 240$ kips $\tag{B-17a}$

550 °C (1,022 °F): $\quad V_p = 0.6 \, (0.5 \times 40)(23)(0.419) = 120$ kips $\tag{B-17b}$

assuming the full web is effective. Hence, the three-bolt capacity is

Room Temperature: $\quad V_u = 3[3(0.75)(0.375)(60)] \cong 152$ kips $\tag{B-18a}$

550 °C (1,022 °F): $\quad V_u = 3[3(0.75)(0.375)(30)] \cong 76$ kips $\tag{B-18b}$

The double shear capacity of the three 3/4-inch high-strength bolts is

Room Temperature: $\quad R_u = 3[2(0.62)(1.0)(1.18 \, F_{ub})(A_b)] \cong 232$ kips $\tag{B-19a}$

550 °C (1,022 °F): $\quad R_u = 3[2(0.62)(0.5)(1.18 \, F_{ub})(A_b)] \cong 116$ kips $\tag{B-19b}$

where F_{ub} = 120 ksi is the bolt tensile strength at room temperature (see Section B.3.3) and A_b = 0.4418 square inch is the bolt area. At 550 °C (1,022 °F), the bolt tensile strength is approximately $0.51(1.18 \, F_{ub}) \cong 71$ ksi.

This verifies that the bolted shear connections have sufficient capacity to develop the reduced plastic moment capacity of the fire-weakened steel beam cantilever and sustain large vertical deformation.

The failures all appear to be a result of the large tensile force that developed in the structural system during the fire and/or as the structure cooled. As demonstrated in Section B.4.2, the tensile capacity of the bolted shear splice in the beam web can be estimated for a bolt as

$$R_u = L_e t \, F_u \leq 3 \, d \, t \, F_u \tag{B-20}$$

with L_e = 1.344 inches is the edge distance, t = 0.375 inch is the plate thickness in bearing, and F_u is the applicable tensile strength.

Room Temperature: $\quad R_u = (1.344)(0.375)(60) = 30$ kips/bolt $\tag{B-21a}$

550 °C (1,022 °F): $\quad R_u = (1.344)(0.375)(30) = 15$ kips/bolt $\tag{B-21b}$

The photographs in Figure 4-22, in Chapter 4, indicate that the deformed structure subjected the bolted shear connection to a large tensile force. At 550 °C (1,022 °F), the ultimate resistance of the three bolts is about 45 kips. The capacity increases to about 90 kips at room temperature. Failure occurred between these bounds.

Tensile catenary action of this type of floor framing members and their connections has not been a design requirement or consideration for most buildings. For the analysis shown here, with assumed fire temperatures, increasing the end distance L_e to 2.25 inches would increase the tensile capacity of the three bolts to about 76 kips at 550 °C (1,022 °F) and 152 kips at room temperature, because the resistance would increase to the limit in Equation B-20.

B.5 References

AISC. 2001. *Manual of Steel Construction*, LRFD. 3rd Edition. Chicago.

ASTM. 1999. ASTM E1820 *Standard Test Method for Measurement of Fracture Toughness*.

Brockenbrough, R.L., and Johnston, B.G. 1968. *USS Steel Design Manual*. USS, Pittsburgh, PA.

Fisher, J. W., Galambos, T. V., Kulak, G. L., and Ravindra, M. K. 1978. "Load and Resistance Factor Design Criteria for Connectors," *Journal of Structural Division*. ASCE, Vol. 104, No. ST9, September.

Kulak, G. L., Fisher, J. W., and Struik, J. H. A. 1987. *Guide to Design Criteria for Bolted and Riveted Joints*. 2nd Edition. Wiley, NY.

Lesik, D. F. and Kennedy, D. J. L. 1990. "Ultimate Strength of Fillet Welded Connections Loaded In-Plane," *Canadian Journal of Civil Engineering*. Vol. 17, No. 1.

PATH-NYNJ. 1976. PATH-NYNJ Document 761101, *The World Trade Center: A Building Project Like No Other*. May.

Salmon, C. G., and Johnson, J. E. 1996. *Steel Structures, Design and Behavior*. 4th Edition. Harper Collins.

Structural Steel Design. 1974. L. Tall, Ed. 2nd Edition. Ronald Press.

USX. 1998. *Making, Shaping, and Treating of Steel*.

Jonathan Barnett
Ronald R. Biederman
R. D. Sisson, Jr.

C Limited Metallurgical Examination

C.1 Introduction

Two structural steel members with unusual erosion patterns were observed in the WTC debris field. The first appeared to be from WTC 7 and the second from either WTC 1 or WTC 2. Samples were taken from these beams and labeled Sample 1 and Sample 2, respectively. A metallurgic examination was conducted.

C.2 Sample 1 (From WTC 7)

Several regions in the section of the beam shown in Figures C-1 and C-2 were examined to determine microstructural changes that occurred in the A36 structural steel as a result of the events of September 11, 2001, and the subsequent fires. Although the exact location of this beam in the building was not known, the severe erosion found in several beams warranted further consideration. In this preliminary study, optical and scanning electron metallography techniques were used to examine the most severely eroded regions as exemplified in the metallurgical mount shown in Figure C-3. Evidence of a severe high temperature corrosion attack on the steel, including oxidation and sulfidation with subsequent intergranular melting, was readily visible in the near-surface microstructure. A liquid eutectic mixture containing primarily iron, oxygen, and sulfur formed during this hot corrosion attack on the steel. This sulfur-rich liquid penetrated preferentially down grain boundaries of the steel, severely weakening the

Figure C-1   Eroded A36 wide-flange beam.

APPENDIX C: *Limited Metallurgical Examination*

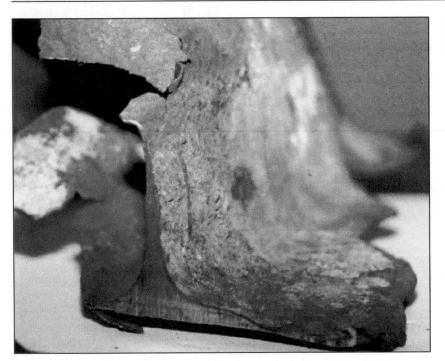

*Figure C-2
Closeup view of eroded wide-flange beam section.*

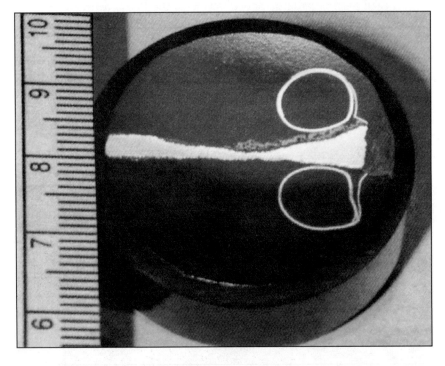

*Figure C-3
Mounted and polished severely thinned section removed from the wide-flange beam shown in Figure C-1.*

beam and making it susceptible to erosion. The eutectic temperature for this mixture strongly suggests that the temperatures in this region of the steel beam approached 1,000 °C (1,800 °F), which is substantially lower than would be expected for melting this steel.

When steel cools below the eutectic temperature, the liquid of eutectic composition transforms to two phases, iron oxide, FeO, and iron sulfide, FeS. The product of this eutectic reaction is a characteristic geometrical arrangement that is unique and is readily visible even in the unetched microstructure of the steel. Figures C-4 and C-5 present typical near-surface regions showing the microstructural changes that occur due to this corrosion attack. Figure C-6 presents the microstructure from the center of a much thicker

APPENDIX C: *Limited Metallurgical Examination*

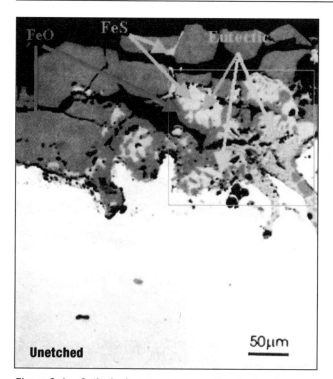

Figure C-4   Optical microstructure near the steel surface.

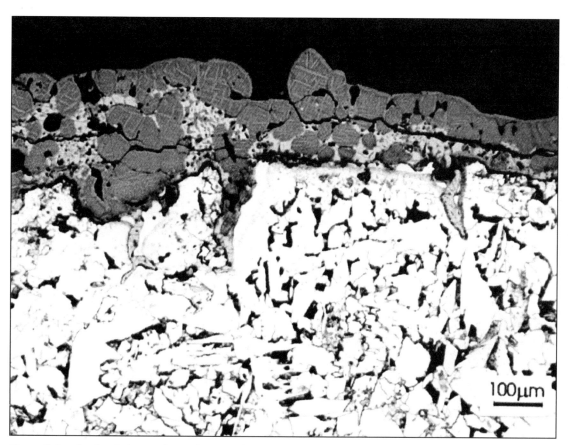

Figure C-5   Another hot corrosion region near the steel surface (etched with 4 percent nital). (Note: (1) Oxide rounding where the FeO-FeS eutectic product is present. (2) Reduction in banding of the steel when re-transformation occurs on cooling from austenite to ferrite and pearlite.)

APPENDIX C: *Limited Metallurgical Examination*

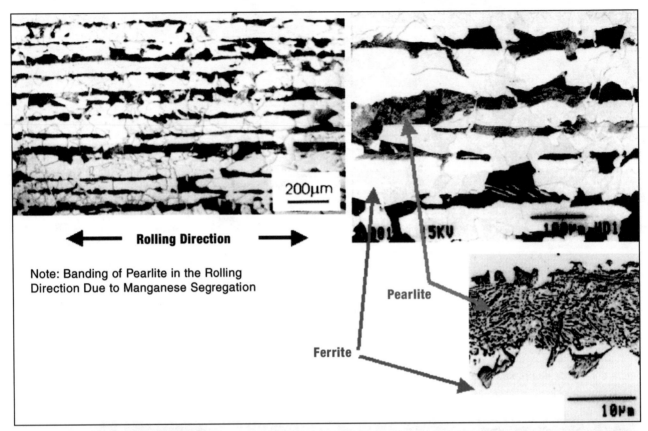

Figure C-6   Microstructure of A36 steel.

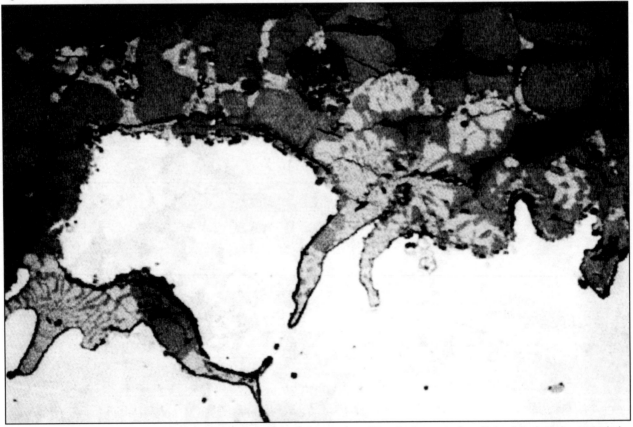

Figure C-7   Deep penetration of liquid into the steel. (Note: Hot corrosion of the steel can produce "islands" of steel surrounded by liquid, which will make erosion of the steel much easier.)

APPENDIX C: *Limited Metallurgical Examination*

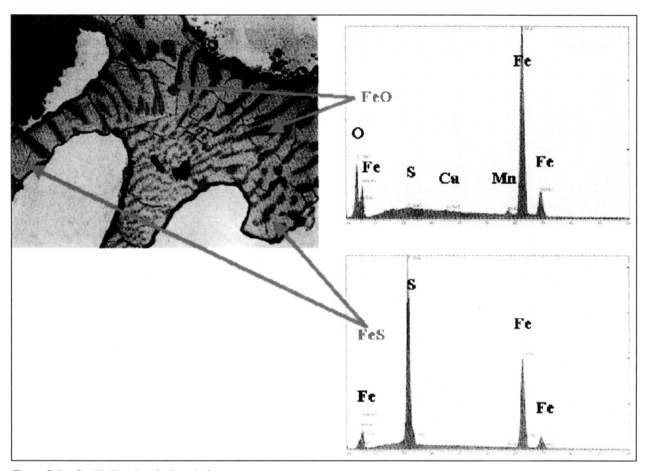

Figure C-8   Qualitative chemical analysis.

section of the steel that is unaffected by the hot corrosion. Figure C-7 illustrates the deep penetration of the liquid into the steel's structure. In order to identify the chemical composition of the eutectic, a qualitative chemical evaluation was done using energy dispersive X-ray analysis (EDX) of the eutectic reaction products. Figure C-8 illustrates the results of this analysis.

C.3   Summary for Sample 1

1. The thinning of the steel occurred by a high-tempertature corrosion due to a combination of oxidation and sulfidation.

2. Heating of the steel into a hot corrosive environment approaching 1,000 °C (1,800 °F) results in the formation of a eutectic mixture of iron, oxygen, and sulfur that liquefied the steel.

3. The sulfidation attack of steel grain boundaries accelerated the corrosion and erosion of the steel.

C.4   Sample 2 (From WTC 1 or WTC 2)

The origin of the steel shown in Figure C-9 is thought to be a high-yield-strength steel removed from a column member. The steel is a high-strength low-alloy (HSLA) steel containing copper. The unusual thinning of the member is most likely due to an attack of the steel by grain boundary penetration of sulfur forming sulfides that contain both iron and copper. Figures C-10, C-11, and C-12 show the region of severe corrosion at different levels of magnification.

APPENDIX C: Limited Metallurgical Examination

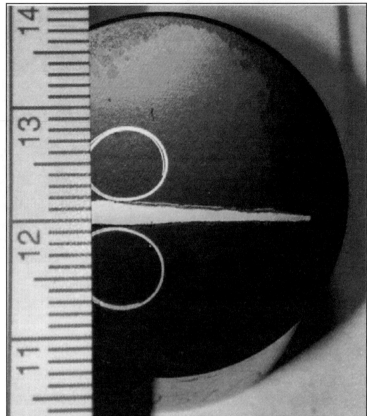

Figure C-9 Qualitative chemical analysis.

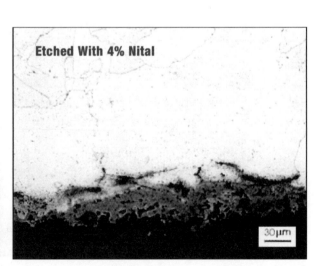

Figure C-10 Grain boundary corrosion attack.

APPENDIX C: *Limited Metallurgical Examination*

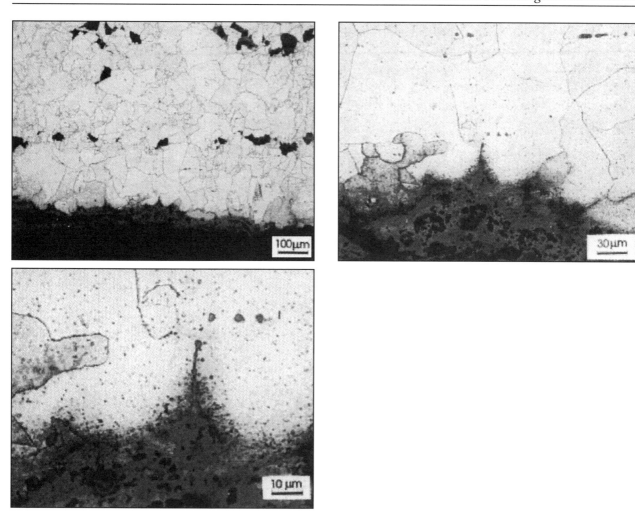

Figure C-11 Microstructure of a typical region showing the surface and grain boundary corrosion attack of Sample 2.

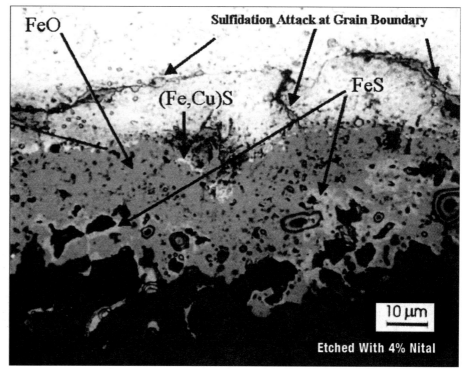

*Figure C-12
Higher magnification of the region shown in Figure C-10.*

APPENDIX C: *Limited Metallurgical Examination*

Figure C-13 shows the region where a qualitative chemical analysis of the eroded region was performed. The comparison of the EDX spectra from the specific regions identified in Figure C-13 shows concentration of copper and sulfur in the grain boundaries in addition to iron sulfide formation adjacent to iron oxide in the oxidized surface layer. Sulfide formation within the steel microstructure increases in concentration as the oxidized region is approached from the steel side. This is clearly shown in Figure C-14.

The larger sulfides further into the steel are the more stable manganese sulfides that were formed when the steel was made. The smaller sulfides that have formed as a result of the fire do not contain significant amounts of manganese, but rather are primarily sulfides containing iron and copper. These sulfides have a lower melting temperature range than manganese sulfide. It is much more difficult to tell if melting has occurred in the grain boundary regions in this steel as was observed in the A36 steel from WTC 7. It is possible and likely, however, that even if grain boundary melting did not occur, substantial penetration by a solid state diffusion mechanism would have occurred as evidenced by the high concentration of sulfides in the grain interiors near the oxide layer. Temperatures in this region of the steel were likely to be in the range of 700–800 °C (1,290–1,470 °F).

Figure C-13 Regions where chemical analysis was performed.

APPENDIX C: *Limited Metallurgical Examination*

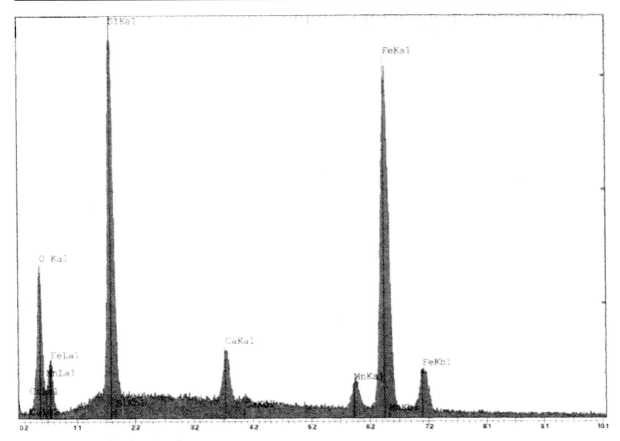

Location 1 (Figure C-13 continued).

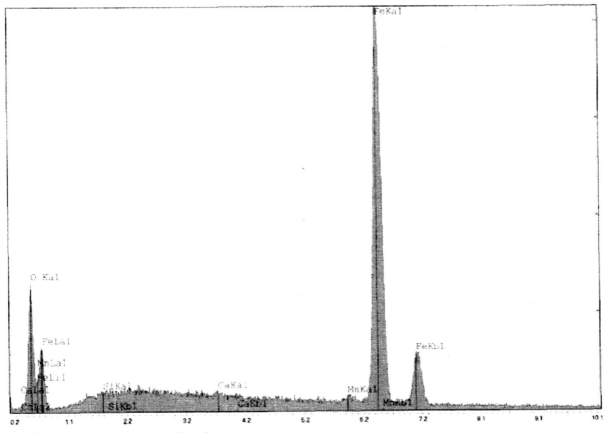

Locations 2 and 3 (Figure C-13 continued).

APPENDIX C: *Limited Metallurgical Examination*

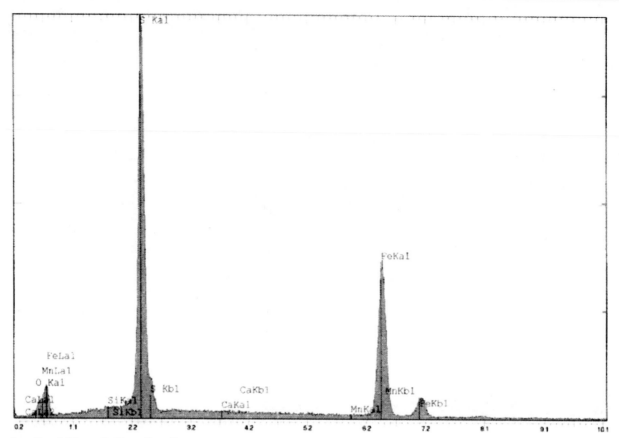

Location 4 (Figure C-13 continued).

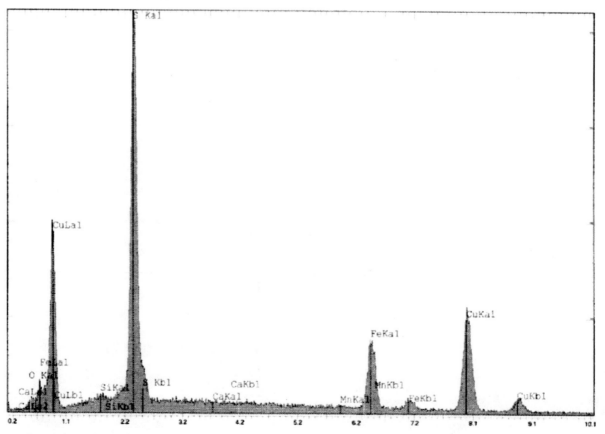

Location 5 (Figure C-13 continued).

APPENDIX C: *Limited Metallurgical Examination*

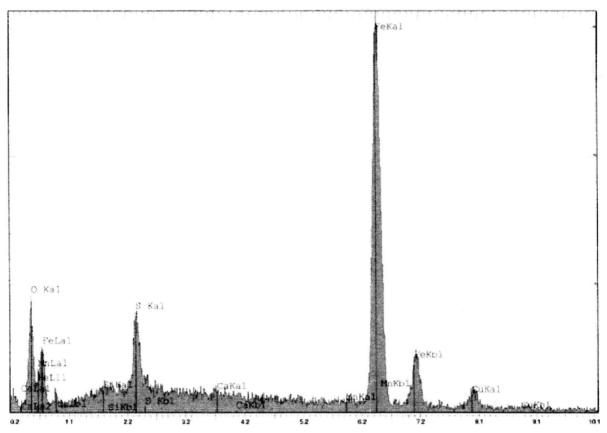

Location 6 (Figure C-13 continued).

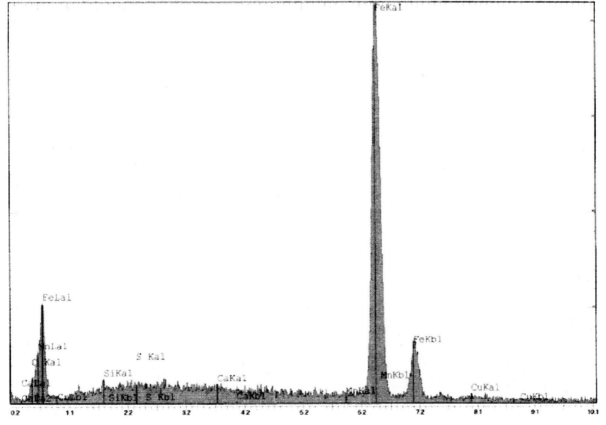

Location 7 (Figure C-13 continued).

FEDERAL EMERGENCY MANAGEMENT AGENCY

APPENDIX C: *Limited Metallurgical Examination*

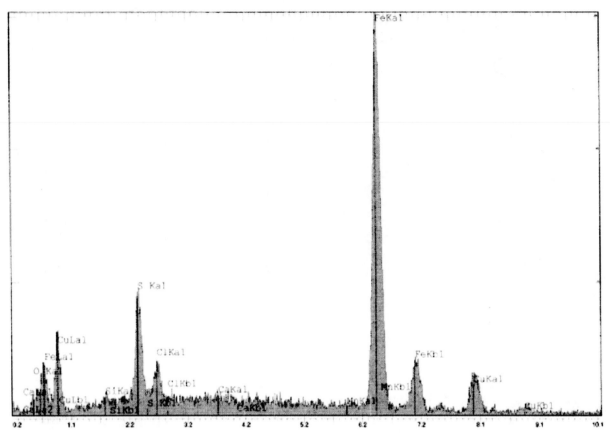

Location 8 (Figure C-13 continued).

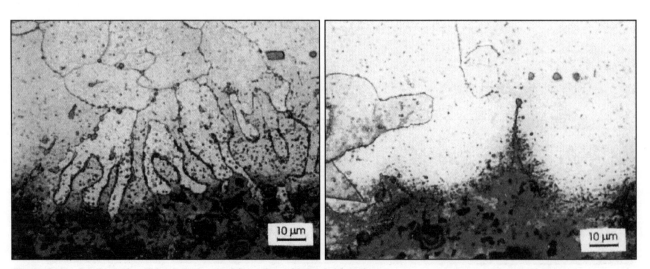

Figure C-14 Gradient of sulfides into the steel from the oxide-metal interface.

C.5 Summary for Sample 2

1. The thinning of the steel occurred by high temperature corrosion due to a combination of oxidation and sulfidation.

2. The sulfidation attack of steel grain boundaries accelerated the corrosion and erosion of the steel.

3. The high concentration of sulfides in the grain boundaries of the corroded regions of the steel occured due to copper diffusing from the HSLA steel combining with iron and sulfur, making both discrete and continuous sulfides in the steel grain boundaries.

C.6 Suggestions for Future Research

The severe corrosion and subsequent erosion of Samples 1 and 2 are a very unusual event. No clear explanation for the source of the sulfur has been identified. The rate of corrosion is also unknown. It is possible that this is the result of long-term heating in the ground following the collapse of the buildings. It is also possible that the phenomenon started prior to collapse and accelerated the weakening of the steel structure. A detailed study into the mechanisms of this phenomenon is needed to determine what risk, if any, is presented to existing steel structures exposed to severe and long-burning fires.

Ramon Gilsanz
Audrey Massa

D WTC Steel Data Collection

D.1 Introduction

WTC steel data collection efforts were undertaken by the Building Performance Study (BPS) Team and the Structural Engineers Association of New York (SEAoNY) to identify significant steel pieces from WTC 1, 2, 5, and 7 for further study. The methods used to identify and document steel pieces are presented, as well as a spreadsheet that documents the data for steel pieces inspected at various sites from October 2001 through March 2002.

D.2 Project Background

Collection and storage of steel members from the WTC site was not part of the BPS Team efforts sponsored by FEMA and the American Society of Civil Engineers (ASCE). SEAoNY offered to organize a volunteer team of SEAoNY engineers to collect certain WTC steel pieces for future building performance studies. Visiting Ground Zero in early October 2001, SEAoNY engineers, with the assistance from the New York City Department of Design and Construction (DDC), identified and set aside some steel pieces for further study.

Of the estimated 1.5 million tons of WTC concrete, steel, and other debris, more than 350,000 tons of steel have been extracted from Ground Zero and barged or trucked to salvage yards where it is cut up for recycling. Salvage yard operations are shown in Figures D-1 through D-3. Four salvage yards were contracted to process WTC steel:

- Hugo Nue Schnitzer at Fresh Kills (FK) Landfill, Staten Island, NJ
- Hugo Nue Schnitzer's Claremont (CM) Terminal in Jersey City, NJ
- Metal Management in Newark (NW), NJ
- Blanford and Co. in Keasbey (KB), NJ

SEAoNY appealed to its membership for experienced senior engineers to visit the salvage yards on a volunteer basis, and to identify and set aside promising steel pieces for further evaluation. Seventeen volunteer SEAoNY engineers started going to the yards in November 2001. A list of engineers and others who contributed to this effort is included in Appendix G of this report.

As of March 15, 2002, a total of 131 engineer visits had been made to these yards on 57 separate days. An engineer visit typically ranged from a few hours to an entire day at a salvage yard. The duration of the

APPENDIX D: *Steel Data Collection Project Summary*

Figure D-1
Mixed, unsorted steel upon delivery to salvage yard.

visits, number of visits per yard, and the dates the yards were visited varied, depending on the volume of steel being processed, the potential significance of the steel pieces being found, salvage yard activities, weather, and other factors. Sixty-two engineer trips were made to Jersey City, 38 to Keasbey, 15 to Fresh Kills, and 16 to Newark. Three trips made in October included several ASCE engineers. Eleven engineer trips were made in November, 41 in December, 43 in January, 28 in February, and 5 through March 15, 2002.

D.3 Methods

Engineers identified steel members that would be considered for evaluation or tests relative to the fire and structural response of the WTC buildings. Pieces that were measured and determined to be significant were marked to be saved, and arrangements were made to have them moved to a safe location where they would not be processed (cut up and shipped). Some pieces were not saved, but samples, called coupons, were cut from them and saved for future studies.

D.3.1 Identifying and Saving Pieces

As shown in Figure D-4, the engineers searched through unsorted piles of steel for pieces from WTC 1 and WTC 2 impact areas and from WTC 5 and WTC 7. They also checked for pieces of steel exposed to fire. Specifically, the engineers looked for the following types of steel members:

- Exterior column trees and interior core columns from WTC 1 and WTC 2 that were exposed to fire and/or impacted by the aircraft.

- Exterior column trees and interior core columns from WTC 1 and WTC 2 that were above the impact zone.

- Badly burnt pieces from WTC 7.

APPENDIX D: *Steel Data Collection Project Summary*

Figure D-2 Torch cutting of very large pieces into more manageable pieces of a few tons each.

Figure D-3 Pile of unsorted, mixed steel (background) with sorted, large steel pieces (center) being lifted and cut into smaller pieces (left).

APPENDIX D: *Steel Data Collection Project Summary*

Figure D-4   Engineer climbing in unprocessed steel pile to inspect and mark promising pieces.

- Connections from WTC 1, 2, and 7, such as seat connections, single shear plates, and column splices.

- Bolts from WTC 1, 2, and 7 that were exposed to fire, fractured, and/or that appeared undamaged.

- Floor trusses, including stiffeners, seats, and other components.

- Any piece that, in the engineer's professional opinion, might be useful for evaluation. When there was any doubt about a particular piece, the piece was kept while more information was gathered. A conservative approach was taken to avoid having important pieces processed in salvage yard operations.

The engineers were able to identify many pieces by their markings. Each piece of steel was originally stenciled in white or yellow with information telling where it came from and where it was going. A sample of the markings can be seen in Figure D-5.

For example, a given piece might be marked, "PONYA WTC 213.00 236B4-9 558 35 TONS." Translated, this meant the column was destined for the Port of New York Authority's World Trade Center as part of contract number 213.00. Its actual piece number was 236B, and it was to be used between floors 4 and 9 in tower B (WTC 2). Its derrick division number was 558, which determined which crane would lift it onto the building and the order in which it was to be erected. Other markings might include the name of the iron works or shipping instructions to those responsible for railway transportation (Gillespie 1999).

Additional markings (and duplicates of stenciled markings) may sometimes be found stamped into the steel pieces. These stamped markings are about 3/4 inch tall.

APPENDIX D: *Steel Data Collection Project Summary*

Figure D-5   Stenciled markings on WTC 2 perimeter column from floors 68–71.

In the absence of markings, member size is the quickest and easiest means for the engineers to establish an approximate original location for a piece. For example, the spandrel plates used in the column-to-column connections in the perimeters of WTC 1 and WTC 2 reportedly ranged in thickness from about 1-1/2 inches at the lower levels to as little as 3/8 inch at the upper levels.

The lighter perimeter columns from WTC 1 and WTC 2 appear to have used column-to-column connections with 4 bolts, whereas larger members presumably from lower floors used six-bolt column-to-column connections. Core column sizes vary, with some heavier sections at the lower floors having plates 4 inches thick or greater.

After a steel piece was identified for further study, it was set aside. As shown in Figure D-6, each piece was marked with spray paint, labeled "SAVE" and a piece number, such as "C-68." The engineers also advised site personnel of the location of these pieces so they would not be processed as scrap.

D.3.2   Documenting Pieces

To document the identified steel pieces of interest, the engineers measured their dimensions. They also drew sketches, and took photographs and videos of the pieces.

The steel member dimensions helped to determine the approximate building location of a piece prior to the disaster. The engineers measured and recorded dimensions using metal tape rules, vernier calipers, or other measuring devices. See Figures D-7 through D-9.

APPENDIX D: *Steel Data Collection Project Summary*

Figure D-6 Steel pieces marked "SAVE."

Figure D-7 Engineers measuring and recording steel piece dimensions.

APPENDIX D: *Steel Data Collection Project Summary*

Figure D-8
Engineer measuring spandrel plate thickness (t_s).

Figure D-9 Measurement of 1/4 inch for web thickness (t_w).

APPENDIX D: *Steel Data Collection Project Summary*

The measured and recorded dimensions (shown in Figure D-10) included the following:

- depth of the piece (d)
- thickness of the web (t_w)
- length of flange (b_f)
- thickness of the top flange (t_{tf})
- thickness of the bottom flange (t_{bf})
- thickness of the spandrel plate (t_s)

Note that the thickness of the spandrel plate may be different from that of the top flange.

D.3.3 Getting Coupons

Samples, or coupons, were cut by yard personnel. A coupon is a sample of steel cut from a larger portion of a steel member or piece. The collected coupons cut are intended for off-site examination in a laboratory.

Where possible, coupons were selected to yield sufficient material for a number of destructive (and mutually exclusive) tests on steel from essentially the same condition. Coupons were sized to be 12 inches by 12 inches, which is considered adequate for most purposes. Where possible, coupons included two faces of attached plates forming a portion of the member. They were also selected so that heat effects from the cutting operation did not affect the coupons' intended test areas.

Figure D-11 shows a steel piece clearly marked with spray paint that shows salvage yard personnel where to cut the coupon. A coupon that has been cut is shown in Figure D-12.

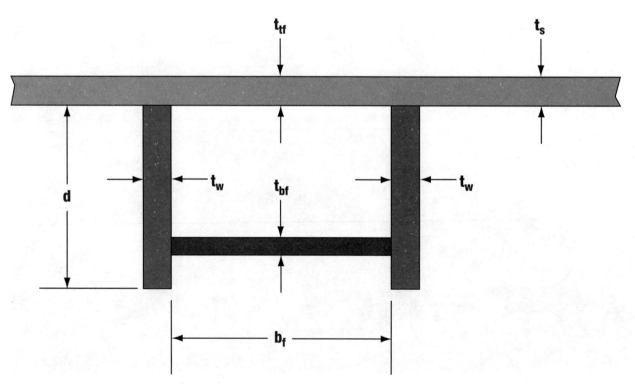

Figure D-10 Measured dimensions of the steel pieces.

APPENDIX D: *Steel Data Collection Project Summary*

Figure D-11 Burnt steel piece marked for cutting of coupon.

Figure D-12 Coupon cut from WTC 5 showing web tear-out at bolts.

APPENDIX D: *Steel Data Collection Project Summary*

D.4 Data Collected

The steel data are compiled in a spreadsheet that includes data from each of the four salvage yards visited by the SEAoNY and WTC BPS Team engineers (the spreadsheet is presented at the end of this appendix). The data are organized according to the salvage yard where each steel piece was examined. The data include the piece identification mark that was sprayed on the piece, the measured dimensions, a brief description of the piece indicating why the piece was selected for further evaluation, information identifying photographs and/or video taken, and the status of any coupon taken. Pieces that were searched for and inspected include perimeter or core columns near the impact area of WTC 1 or WTC 2, burnt pieces from WTC 7, and connection pieces from WTC 5 (see Figures D-12 through D-18).

The steel pieces range in size from fasteners inches in length and weighing a couple of ounces to column pieces up to 36 feet long and weighing several tons. As of March 15, 2002, a total of 156 steel pieces (not including most of the fasteners and other smaller pieces) had been inspected. In addition, seven pieces were set aside from Ground Zero with assistance from the DDC.

It is important to note that the quality of the pieces, rather than the number of pieces, is significant to this study. Not all of these pieces were kept for further study. This is because:

- some pieces were later determined not to be relevant to understanding building damage;
- once a coupon was taken, the full piece was discarded; and
- pieces were accidentally processed in salvage yard operations before they were removed from the yards for further study.

Figure D-13
WTC 1 or WTC 2 core column (C-74).

APPENDIX D: *Steel Data Collection Project Summary*

Figure D-14 WTC 7 W14 column tree with beams attached to two floors.

*Figure D-15
Built-up member with failure along stitch welding.*

APPENDIX D: *Steel Data Collection Project Summary*

Figure D-16 *Engineer inspecting fire damage of perimeter column tree from WTC 1 or WTC 2.*

Figure D-17 *Seat connection in fire-damaged W14 column from WTC 7.*

Figure D-18 WTC 1 or WTC 2 floor-truss section with seat connection fractured along welds.

It was expected that most steel members from the impact zones would have reached the yards early in the WTC site excavation process because pieces from the higher floors would be removed first from the debris at Ground Zero. However, barges of steel that were being unloaded in February and March at the Jersey City and Newark salvage yards were found to have pieces from the higher floors.

D.5 Conclusions and Future Work

The ongoing volunteer effort of the SEAoNY engineers is securing WTC steel pieces that will provide physical evidence for studies on WTC building performance. As of March 15, 2002, seventeen engineers, visiting four salvage yards, have identified approximately 150 pieces. Pieces have been identified that are from WTC 1, 2, 5, and 7. Documentary photographs and videos have been taken and coupons collected.

Future studies are expected based on the pieces and data collected. Coupons have been collected for metallurgical tests to determine the temperatures to which they were subjected and their steel characteristics. The National Institute of Standards and Technology (NIST) is currently conducting environmental tests, abating asbestos as necessary, and shipping available pieces to its Gaithersburg, MD, facility for storage and further study. As of May 2002, a total of 41 steel pieces had been shipped to NIST.

D.6 References

Gillespie, A.K. 1999. *Twin Towers, 1999, The Life of New York City's World Trade Center*. Rutgers University Press, New Brunswick, NJ. ISBN 0-8135-2742-2.

Steel Data Collection Spreadsheet

SEAoNY Summary of Identified WTC Steel Pieces at Salvage Yards as of March 15, 2002

> **Note:** As of May 2002, of the 156 steel pieces listed in the spreadsheet, 41 are at the National Institute of Standards, 19 were discarded after coupons were taken, 45 are at the salvage yards, and the rest either were discarded after they were documented or were accidentally processed in the salvage operation before or after being documented.

SEAoNY Summary of Identified WTC Steel Pieces at Salvage Yards as of March 15, 2002

	Yard	Trip Date	Note Taker	Piece ID Mark	d	t_w	b_f	t_{tf}	t_{bf}	t_s	~L	Remarks (description, photos/videos, coupons)
1	CM	10/10/01	-	01				See Remarks				One of 10 full-size pieces of exterior and interior columns marked during ASCE visits. These pieces were accidentially processed in salvage yard operations before being documented.
2	CM	10/10/01	-	02	13-1/2"	2-1/2"	9"	7/8"	7/8"	15/16"	36'	One full-length perimeter column w/2 99" long spandrel plates suggesting use in sky lobby location. Piece painted with markings "ASTANEH 2 ASCE". Piece was in Astaneh's pile and was measured by David Sharp of SEAoNY on 3/11/02.
3	CM	10/10/01	-	03	13-1/2"	1-7/8"	10-1/2"	5/8"	5/8"	~3/4"	17'	One perimeter column w/portion of a spandrel plate. Stamped markings "L2042 1 144 45". Piece painted with markings "ASTANEH 3 ASCE". Piece was in Astaneh's pile and was measured by David Sharp of SEAoNY on 3/11/02.
4	CM	10/10/01	-	04				See Remarks				Same as Piece 01, above.
5	CM	10/10/01	-	05				See Remarks				Same as Piece 01, above.
6	CM	10/10/01	-	06				See Remarks				Same as Piece 01, above.
7	CM	10/10/01	-	07				See Remarks				Same as Piece 01, above.
8	CM	10/10/01	-	08				See Remarks				Same as Piece 01, above.
9	CM	10/10/01	-	09				See Remarks				Same as Piece 01, above.
10	CM	10/10/01	-	10				See Remarks				Same as Piece 01, above.
11	CM	10/10/01	-	11				See Remarks				Fire-damaged piece of wide flange. Otherwise, same as Piece 01, above.
12	CM	12/19/01	Gilsanz	AA	11"	1-3/8"	11-1/4"	9/16"	n/a	1-1/4"		(Photo:DKoutsoubis:1/30/02:#43-48) From WTC 1 or 2 with fire damage &/or floor location. Depth measurement does not include portion of web beyond bottom flange.
13	CM	12/19/01	Gilsanz	BB	11"	n/a	11-3/8"	7/16"	n/a	9/16"		(Photo:DKoutsoubis:1/30/02:#49-52). From WTC 1 or 2 with fire damage &/or floor location. Depth measurement does not include portion of web beyond bottom flange.
14	CM	12/19/01	Gilsanz	CC	10-7/8"	1-1/8"	11-1/2"	7/16"	3/8"	9/16"		(Photo:DKoutsoubis:1/30/02:#53-58) From WTC 1 or 2 with fire damage &/or floor location. Depth measurement does not include portion of web beyond bottom flange.

SEAoNY Summary of Identified WTC Steel Pieces at Salvage Yards as of March 15, 2002

	Yard	Trip Date	Note Taker	Piece ID Mark	d	t_w	b_f	t_{ff}	t_{bf}	t_s	~L	Remarks (description, photos/videos, coupons)
15	CM	12/19/01	Gilsanz	DD	13-7/8"	1-1/8"	11-1/4"	1/4"	3/8"	1/2"		(Photo:DKoutsoubis:1/30/02:#59-62) From WTC 1 or 2 with fire damage &/or floor location.
16	CM	12/19/01	Gilsanz	EE	n/a	7/16"	n/a	~1/4"	~1/4"	7/16"		(Photo:DKoutsoubis:1/30/02:#74-82) From WTC 1 or 2 with fire damage &/or floor location.
17	CM	12/19/01	Sharp	FF	13-1/2"	1-9/16"	10-5/8"	~3/8"	7/16"	1-1/4"	25'	Piece marked 12/19/01, measured 1/30/02. (Photo:DKoutsoubis:1/30/02:#83-88) Single bow-shaped badly burned perimeter column. "Left" column from interior perspective determined by bolt splice. Primer gone.
18	CM	12/19/01	Sharp	GG				See Remarks				Piece marked 12/19/01, measured 1/30/02. Sample of a bolt splice taken from a perimeter column from sky-lobby floor as indicated by two large spandrel plates. Piece was selected for intact bolt splice only. Photo:DKoutsoubis:1/30/02:#89-97.
19	CM	12/19/01	Sharp	HH	12"	1/2"	12-1/4"	3/4"	3/4"	n/a	14'	Piece marked 12/19/01, measured 1/30/02. (Photo:DKoutsoubis:1/30/02:#98-102) Core column section with bottom and splice intact, brittle failure in midpoint.
20	CM	12/19/01	Gilsanz / Sharp	A or W14A	16"		15-1/2"	1-3/16"				WTC#7 W14 chosen for seat section response. Seat measurements taken. Seat is 8" X 7-1/2" X 5-1/4" X 3/8" thick. Stiffener is 9-5/8" X 5-1/4" X 3/8" thick. Bolt holes are 15/16". Photo by DKoutsoubis:1/30/02:#68-73.
21	CM	12/19/01	Gilsanz / Sharp	B or W14B			See Remarks					WTC#7 W14 chosen due to seat section response. Seat measurements taken. Seat is 7-1/2" X 7-1/4" X 3/8" thick. Photo by DKoutsoubis:1/30/02:#63-67.
22	CM	1/0802	Bonilla	C10	11-1/2"	3/16"	14"	3/8"	5/16"	3/8"		Chosen due to floor location. Perimeter column from WTC 1 or 2. Piece marked "88-85 L2337 451". DSharp photo.
23	CM	1/8/02	Bonilla	C11	11-1/2"	n/a	14"	1/4"	1/4"	3/8"		Chosen due to floor location. Perimeter column from WTC 1 or 2. DSharp photo.
24	CM	1/21/02	Koutsoubis	C12a	14-3/4"	5/8"	14-3/4"	1"				Chosen due to connection type. WTC#7 W14 tree column-column piece. (Photo:DKoutsoubis:1/21/02:#40-45)
	CM	1/21/02	Koutsoubis	C12b	23-5/8"	1/2"	7"	5/8"				WTC#7 W14 tree column-beam piece. Beam arm length is 4' 5-1/8" Attached to C12a. (Photo:DKoutsoubis:1/21/02:#40-45).
25	CM	1/23/02	Sharp	C13a	13-3/4"	1/4"	13-1/2"	1/4"	1/4"	3/8"	~30 ft	Chosen due to floor location. Markings on Steel: "321 B200 92 90 3T <569>". Possibly indicates f/corner of WTC 1 or 2. (Photo:DKoutsoubis:1/25/02:#15-19).
	CM	1/23/02	Sharp	C13b	7-5/8"	1/4"	12-1/2"	5/16"	1/4"	3/8"		Box column attached to C-13a. (Photo:DKoutsoubis:1/25/02:#20-25).
26	CM	1/23/02	Sharp	C14a	13-1/2"	1/4"	13-1/4"	5/16"	5/16"	3/8"		Chosen due to floor location. Perimeter column from WTC 1 or 2. (Photo:DKoutsoubis:1/25/02:#10-14) Markings "3T <570> B300 85-87".

SEAoNY Summary of Identified WTC Steel Pieces at Salvage Yards as of March 15, 2002

	Yard	Trip Date	Note Taker	Piece ID Mark	d	t_w	b_f	t_{ff}	t_{bf}	t_s	~L	Remarks (description, photos/videos, coupons)
	CM	1/24/02	Sharp	C14b	n/a	5/16"	5-7/8"	n/a	n/a	n/a		Small portion of a box section, attached to C14a. Small 'Tin' box too.
27	CM	1/23/02	Sharp	C15	13-3/8"	1/4"	13-1/4"	1/4"	~3/8"	3/8"		Perimeter Column from WTC 1 or 2, no clearly evident markings. Chosen due to floor location.(Photo:DKoutsoubis:1/25/02:#26-30).
28	CM	1/23/02	Sharp	C16	13-3/4"	1/4"	13"+	1/4"	1/4"	3/8"	~15ft	Marking on t+M43op of diaphragm plate (or top of column?) "75 1 203" Chosen due to floor location. (Photo:DKoutsoubis:1/25/02:#90-94).
29	CM	1/23/02	Sharp	C17	13-1/2"	1/4"	13-1/2"	1/4"	3/16"	3/8"		Two perimeter columns from WTC 2?. Chosen due to floor location. Markings: "92-95" "248B 224" (Photo:DKoutsoubis:1/25/02:#39-42).
30	CM	1/23/02	Sharp	C18	13-3/4"	1/4"	13-3/8"	1/4"	1/4"	3/8"		Nut and washer found in box, as well as a letter from AON "Two World Trade Center". (Photo:DKoutsoubis:1/25/02:#31-37) Markings: "+3/16"
31	CM	1/23/02	Sharp	C19	13-1/2"	1/4"	12-3/4"	1/4"	1/4"	3/8"	~15ft	Single perimeter column from WTC 1 or 2. Chosen due to floor location. Fire damage. (Photo:DKoutsoubis:43-48).
32	CM	1/23/02	Sharp	C20	13-1/4"	1/4"	13-3/8"	5/16"	1/4"	3/8"		Three perimeter columns together, chosen due to floor location and some evidence of fire damage on connection. (Photo:DKoutsoubis:1/25/02:#51-55) Markings "4T <557> 91-94" Floor connection upside down?
33	CM	1/23/02	Bonilla	C21	13-1/2"	1/4"	13-1/2"	1/4"	1/4"	3/8"		Perimeter Column from WTC 1 or 2 Chosen due to floor location. (Photo:DKoutsoubis:1/25/02:#65-70).
34	CM	1/23/02	Bonilla	C22	13-1/2"	1/4"	13-1/4"	1/4"	1/4"	3/8"		Piece is marked: "PONYA 4T <69> A?? 93 - ??. Chosen due to floor location and moderate fire damage. Stamped "3145 96-93" (Photo:DKoutsoubis:1/25/02:#124-127).
35	CM	1/23/02	Bonilla	C23	13-1/2"	9/32"	13-3/8"	5/16"	1/4"	3/8"		Perimeter column section flared apart, no fire, column ends peeled apart. Chosen due to floor location.
36	CM	1/25/02	Bonilla	C24	13-1/2"	5/8"	13"	5/16"	1/4"	3/8"		(Photo:DKoutsoubis:1/25/02:#83-86) Chosen due to availability of column to column bolts being intact, 9 of 12 3/4" bolts available Depth of spandrel 8' 3". Piece marked "77-78 203".
37	CM	1/25/02	Bonilla	C25	13-1/2"	1/4"+	13"	5/16"	1/4"	5/8"	~16 ft	(Photo:DKoutsoubis:1/25/02:#83-86) Chosen due to floor location. No clearly evident fire damage, single perimeter column w/2 spandrel plates Crayon markings: "A206 <69> 89-92".
38	CM	1/25/02	Sharp	C26	27-1/4"	1/2"	10-1/4"	3/4"	3/4"	3/8"		(Photo:DKoutsoubis:1/25/02:#56-64) Chosen due member type, 2 1" gusset plates w/3 connecting W sections. Main section has utility cut-outs RB&W ASTM A490 bolts.
39	CM	1/25/02	Sharp	C27	12-1/4"	3/8"	10"	9/16"	9/16"		~14 ft	(Photo:DKoutsoubis:1/25/02:#71-77) Chosen due member type, W12 bow-shaped.

SEAoNY Summary of Identified WTC Steel Pieces at Salvage Yards as of March 15, 2002

	Yard	Trip Date	Note Taker	Piece ID Mark	d	t_w	b_f	t_{tf}	t_{bf}	t_s	~L	Remarks (description, photos/videos, coupons)
40	CM	1/25/02	Bonilla	C28a	13-1/2	5/16"	13"	5/16"	n/a	3/8"		(Photo:DKoutsoubis:1/25/02:#136-139) Chosen due to floor location.
	CM	1/26/02	Bonilla	C28b	7-3/8"	3/8"	12-1/4"	5/16"	n/a	3/8"		Attached to C28a, square column w/rectangular column. Intact seat, 3 or 4 bolts present.
41	CM	1/25/02	Sharp	C29	14-3/4"	1-1/8"	12-1/2"	1-3/4"	1-3/4"			(Photo:DKoutsoubis:1/25/02:#101-107) Chosen due to member type and weight, W14 'C'-shaped.
42	CM	1/25/02	Sharp	C30	16-1/2"	1-1/8"	15-7/8"	1-7/8"	1-7/8"			(Photo:DKoutsoubis:1/25/02:#108-111) Chosen due to member type and weight, W14 'C'-shaped.
43	CM	1/25/02	Koutsoubis	C31a	15-3/8"	13/16"	15-1/2"	1-1/4"	1-1/4"		~24 ft	(Photo:DKoutsoubis:1/25/02:#118-123) W14 Tree. Chosen due to piece type.
	CM	1/25/02	Koutsoubis	C31b	24"	1/2"	7"	5/8"	5/8"			Matching beam attached to C31a column tree.
44	CM	1/25/02	Koutsoubis	C32	14"	1-3/8"	11-1/4"	7/16"	7/16"	11/16"		(Photo:DKoutsoubis:1/25/02:#128-131) Selected by Steficek due to fire damage.
45	CM	1/25/02	Koutsoubis	C33	13-3/4"	1/2"	n/a	1/4"	n/a	7/16"		Selected by Steficek. Chosen due to piece type/original floor location. Photo by DKoutsoubis:1/25/02:#132-135.
46	CM	1/25/02	Koutsoubis	C34	14"	1"	12"	3/8"	3/8"	1/2"		Selected by Steficek. Chosen due to piece type/original floor location. Photo by DKoutsoubis:1/25/02:#140-143.
47	CM	1/25/02	Koutsoubis	C35	14"	7/8"	12-1/2"	1-1/2"	1-1/2"		~36 ft.	(Photo:DKoutsoubis:1/25/02:#144-146) Selected by Bonilla. W section column w/ slot connections.
48	CM	1/30/02	Sharp	C40	13-1/2"	5/16"	13"	1/4"	1/4"	3/8"	~20 ft	(Photo:DKoutsoubis:1/30/02:#6-12) 2 joined perimeter columns chosen due to floor location. Bent in middle. Marking: "98-101 <69> 5T 101".
49	CM	1/30/02	Sharp	C41	13-1/2"	3/8"	13-1/8"	1/4"	n/a	3/8"	~20 ft	(Photo:DKoutsoubis:1/30/02:#13-19) 1 perimeter column with 2 areas of fire damage. Chosen due to floor location.
50	CM	1/30/02	Sharp	C42a	14-5/8"		14-5/8"	1"				(Photo:DKoutsoubis:1/30/02:#20-26) Chosen to member type, column tree W14 w/2 floor connections.
	CM	1/30/02	Sharp	C42b	23-7/8"		7"	9/16"				Beam attached to C42b column.

SEAoNY Summary of Identified WTC Steel Pieces at Salvage Yards as of March 15, 2002

	Yard	Trip Date	Note Taker	Piece ID Mark	d	t_w	b_f	t_{tf}	t_{bf}	t_s	~L	Remarks (description, photos/videos, coupons)
51	CM	1/30/02	Sharp	C43	13"	3/8"	13-1/2"	7/16"	5/16"	3/8"	~20 ft	(Photo:DKoutsoubis:1/30/02:#27-31) Chosen due to floor location. Fire damaged mangled end.
52	CM	1/30/02	Sharp	C44	12-1/8"		10"	5/8"			~13 ft	(Photo:DKoutsoubis:1/30/02:#32-35) W12 X58 Core column? Markings: "<563> 59(S)".
53	CM	1/30/02	Sharp	C45	11-3/4"		12"	5/8"	3/8"		~15 ft	(Photo:DKoutsoubis:1/30/02:#36-42) W12 x65 Core column? Markings: "<563> 16(S2)".
54	CM	2/4/02	Sharp	C46	13-1/2"	5/8"	12-5/8"	1/4"	3/8"	9/16"		(Photo DSharp) 3 column section from tower 2, dimensions taken from 'right' column. Chosen due to location just below impact area Tower B. Markings "PONYA 8T <569> B157-68-71" Moderate fire damage.
55	CM	2/4/02	Sharp	C47	13-1/2"	1/4"	13"	1/4"	1/4"	3/8"	~20 ft	(Photo DSharp) 3 column section. Primer burned off. Rust. Webs split.
56	CM	2/4/02	Sharp	C48	13-1/2"	1/4"	13-1/4"	1/4"	1/4"	3/8"		(Photo DSharp) 2 perimeter columns from 1 or 2, fire damage on bottom.
57	CM	2/4/02	Sharp	C49	13-1/2"	1/4"	12-3/4"	1/4"	1/4"	3/8"	~12 ft	(Photo DSharp) 1 perimeter column w/fire damage.
58	CM	2/4/02	Sharp	C50	13-1/2"	3/8"	13-1/8"	1/4"	1/4"	3/8"		(Photo DSharp) Folded in two. Markings "A103-8?" Piece stamped "L9863" "122-36 100".
59	CM	2/4/02	Sharp	C51	13-1/4"	3/8"	13"	1/4"	1/4"	3/8"		(Photo DSharp) 2 perimeter columns w/90 deg bend on one & 1/2 missing on the other.
60	CM	2/4/02	Sharp	C52	13-1/4"	5/16"	13-1/4"	1/4"	1/4"	3/8"	~15 ft	(Photo DSharp) perimeter column, buckled at spandrel where there is fire damage Unique seat configuration detail.
61	CM	2/4/02	Bonilla	C53	n/a	n/a	n/a	n/a	n/a	n/a		Listed dimensions not applicable.Floor truss with L4" X 6-1/2" Seat: 8 L2x1-1/2 x 1/4. Bolt= 1-1/8" nom dia. Photo DSharp.
62	CM	2/6/02	Sharp	C54	13-1/2"	1/4"	13-1/2"	1/4"	1/4"	none	~4 ft	(Photo DSharp) 1 3 ft piece of a perimeter column w/one plate extending ~15 ft.
63	CM	2/6/02	Sharp	C55	13-1/2"	5/16"	13-1/2"	5/16"	5/16"	none	~8 ft	1 perimeter column section. DSharp photo.
64	CM	2/13/02	Bonilla	C60	10-1/2"	1/2"	12"	3/4"				W Section 2 extreme bends. 'S'-shaped.

SEAoNY Summary of Identified WTC Steel Pieces at Salvage Yards as of March 15, 2002

	Yard	Trip Date	Note Taker	Piece ID Mark	d	t_w	b_f	t_{tf}	t_{bf}	t_s	~L	Remarks (description, photos/videos, coupons)
65	CM	2/13/02	Bonilla	C61	10-3/4"	1/2"	12-1/4"	3/4"				W Section. Markings "150 (S) <69> 13 152W093 367". DSharp photo.
66	CM	2/13/02	Bonilla	C62	11"	1/2"	12-1/4"	5/8"				W Section. Markings "224(S) <48> 7409 351 F450" DSharp photo.
67	CM	2/13/02	Bonilla	C63	14-1/2"	3/8"	13"	1/4"	1/4"	3/8"		Perimeter column markings "97-100" Significant fire damage. 4 fasteners recovered from 'box'. 3 nuts w/bolt piece. 1 nut with threads stripped. DSharp photo.
68	CM	2/13/02	Bonilla	C64	11-1/2"+	1/4"	n/a	1/4"	n/a	3/8"	~12 ft	Perimeter column with only 3 sides remaining. DSharp photo.
69	CM	2/13/02	Bonilla	C65	10-3/4"	1"	12-1/2"	1-1/2"				W Section. 'C'-shaped. Concave side w/slotted clip. Convex side has connection plate. DSharp photo.
70	CM	2/19/02	Massa	C66a	13"	3/4"	13-1/2"	1-1/4"	n/a	n/a		Column Tree -column piece. DSharp photo.
	CM	2/19/02	Massa	C66b	~14-1/2"	5/8"	14-3/4"	1"	n/a	n/a		Column tree -beam piece attached to C66a. DSharp photo.
71	CM	2/19/02	Massa	C67	~11"+	1/4"	~13-1/2"	n/a	1/4"	3/8"		Single perimeter column from WTC 1 or 2 w/spandrel plate. DSharp photo.
72	CM	2/19/02	Massa	C68	13-1/2"	7/16"	13"	1/4"	1/4"	3/8"	~16 ft	4 bolts intact at col-col bearing plate.
73	CM	2/19/02	Massa	C69	16"	3/4"	12"	1-1/2"	n/a	n/a	~7 ft	Double 'T'-section bolted at web. Cracked and missing washers. DSharp photo.
74	CM	2/19/02	Massa	C70	11"	1"	12-1/2"	1-3/4"	n/a	n/a	~10 ft	S'-shaped W-Section w/2 seats. DSharp photo.
75	CM	2/19/02	Massa	C71	11"	1"	12-3/4"	1-3/4"	n/a	n/a	~14 ft	C'-shaped W-Section. DSharp photo.
76	CM	2/19/02	Massa	C72a	12"	3/4"	14-1/2"	1"	n/a	n/a		Column tree w/2 floor connections. Column piece markings "[6] 87-259". DSharp photo.
	CM	2/19/02	Massa	C72b	21-1/2"	~1/2"	~7"	5/8"	n/a	n/a		Column tree w/2 floor connections. Beam piece attached to C72a. DSharp photo.

SEAoNY Summary of Identified WTC Steel Pieces at Salvage Yards as of March 15, 2002

	Yard	Trip Date	Note Taker	Piece ID Mark	d	t_w	b_f	t_{tf}	t_{bf}	t_s	~L	Remarks (description, photos/videos, coupons)
77	CM	2/19/02	Massa	C73	13-1/4"	3/8"	13-1/4"	1/4"	1/4"	3/8"	~15 ft	Perimeter column from WTC 1 or 2 w/spandrel plate. DSharp photo.
78	CM	2/19/02	Massa	C74a	12-1/2"	3/4"	14-1/2"	1"	n/a	n/a	~10 ft	Column tree — column piece attached to C72b. DSharp photo.
	CM	2/19/02	Massa	C74b	~22-1/2"	7/16"	7"	1/2"	n/a	n/a		Column tree — beam piece attached to C74a.
79	CM	2/19/02	Massa	C75	13-1/2"	7/8"	12-1/4"	5/16"	5/16"	3/8"	~15 ft	One perimeter column for WTC 1 or 2 w/spandrel plate. DSharp photo.
80	CM	2/19/02	Massa	C76	18-1/4"	5/8"	12"	2" (1"X2)	n/a	n/a	~13 ft	W-Section w/top and bottom plates splayed apart. Interm. Welds. DSharp photo.
81	CM	2/19/02	Massa	C77	13-1/2"	1/4"	13-1/2"	1/4"	1/4"	3/8"		2 perimeter columns pieces from WTC 1 or 2. Splice at spandrel. DSharp photo.
82	CM	2/19/02	Massa	C78a	12-3/4"	1-3/8"	16"	2-1/8"	n/a	n/a	~14 ft	Column tree, one floor — column piece. DSharp photo.
	CM	2/19/02	Massa	C78b	22"	7/16"	7"	5/8"	n/a	n/a		Column tree, one floor — beam piece attached to C78a. DSharp photo.
83	CM	2/19/02	Massa	C79	4"	3/8"	16"	3/4"	n/a	n/a	~14 ft	Rectangular column, core style from 1 or 2 WTC. DSharp photo."
84	CM	2/25/02	Sharp	C80		15"	7/8"	15-1/2"	1-3/8"	n/a	n/a	Perimeter column marked "603A 92-95 <51>".
85	CM	2/19/02	Massa	C81	~31-1/2"	7/8"	16-1/2"	1-5/8"	1-5/8"	n/a	12'	Photos AMassa and DSharp. Beam section painted "SAVE FEMA G". Flame cut on one end, ductile buckling on the other end.
86	CM	3/11/02	Sharp	C82	? Folded	7/16"	10"	11/16"	2"	n/a	16'	Beam section painted "SAVE FEMA H". Mechanical cut-outs, built-up bottom flange, ductile web buckling. Photo DSharp.
87	CM	3/11/02	Sharp	C83	13"	1-1/2"	34"	1-1/2"	1-1/2"	n/a	23'	Core column box section originally chosen by Astaneh, and placed in his pile. Document by David Sharp of SEAoNY for information purposes.
88	CM	3/11/02	Sharp	C84	13-1/2"	2-5/8"	9-5/8"	7/8"	7/8"	1-1/4"	36'	Piece originally selected by Astaneh and in his pile. Single full length perimeter column with 3 spandrel plates. Bolt splice suggests outer column of tree. Documented by David Sharp of SEAoNY for information purposes.

SEAoNY Summary of Identified WTC Steel Pieces at Salvage Yards as of March 15, 2002

	Yard	Trip Date	Note Taker	Piece ID Mark	d	t_w	b_f	t_{tf}	t_{bf}	t_s	~L	Remarks (description, photos/videos, coupons)
89	CM	3/11/02	Sharp	C85a	12-3/4"	1-3/16"	16"	1-3/4"	1-3/4"	n/a		Piece originally selected by Astaneh and in his pile. Column tree with 9+ floor separation. Full width welded seat. 2 floors+ in length. Documented by David Sharp of SEAoNY for information purposes. Dimensions for column section.
	CM	3/11/02	Sharp	C85b	15"	7/16"	~7"	3/4"	3/4"	n/a		Piece originally selected by Astaneh and in his pile. Beam section from C85a column tree. Documented by David Sharp of SEAoNY for information purposes.
90	CM	3/11/02	Sharp	C86	n/a24"+	1-1/2"	13"	1-1/2"	1-1/2"	n/a	3'	Small folded box section from core. Torch cut, originally chosen by Astaneh and in his pile. Documented by David Sharp of SEAoNY for information purposes.
91	CM	3/11/02	Sharp	C87	~22"	7/16"	10"	1/2"	2"	n/a	25'	Beam section with viscous damper attached. Mechanical cut-outs and built-up bottom flange.
92	CM	3/11/02	Sharp	C88	12"	1-1/2"	16"	1-1/2"	1-1/2"	n/a		Core box column w/two straps and seat, painted "SAVE PA". Also has markings "<557>" Completely splayed apart above seat.
93	CM	3/11/02	Sharp	C89	13-1/2"	1-13/16"	10-1/4"	5/8"	5/8"	1-3/8"	36'	Two badly burned perimeter columns, full length. Spandrel plates flame cut. Markings stamped on piece: "L3 144 146 50 B215 15 12". Piece originally chosen by Astaneh and in his pile. Documented by David Sharp of SEAoNY for information purposes only.
94	CM	3/11/02	Sharp	C90	8-3/4"	3-1/4"	37"	3-1/4"	3-1/4"	n/a	36'	Core column box section originally chosen by Astaneh, and placed in his pile. Documented by David Sharp of SEAoNY for information purposes. Piece is 'S'-shaped and has cracks running along the welds in buckles areas.
95	CM	3/11/02	Sharp	C91	10-5/8"	3/4"	4-1/8"	1-3/8"	1-3/8"	n/a	14'	Beam section with built-up flanges and two seats. Probable core section floor truss connection. Splice plates on both ends.
96	CM	3/14/02	Sharp	C92	13-1/2"	~1/2"	13"	1/4"	1/4"	3/8"	15'	Single perimeter column with 2 seats. 'Outside' column from 'right' via view from building interior. (bolt splice on right).
97	CM	3/14/02	Sharp	C93	13-1/2"	5/16"	13-1/4"	1/4"	1/4"	3/8"	8'	One portion of a light construction perimeter column.
98	KB	11/20/01	Gilsanz/Steficek				See Remarks					Portion of spandrel plate and two bolts from WTC 1 or 2. Coupon A. Photos: A-1(coupon cut location), A-2 (coupon side and top view), A-3(coupon bottom). Steficek video.
99	KB	11/20/01	Gilsanz/Steficek				See Remarks					Coupon B of web and flange sections from perimeter column from WTC 1 or 2. Photos: B-1(coupon cut location), B-2 (coupon front), B-3(coupon perpendicular view). Steficek video
100	KB	11/20/01	Gilsanz/Steficek				See Remarks					Coupon C of web and flange sections from perimeter column from WTC 1 or 2. Photos: C-1(coupon cut location), C-2 (coupon full view), C-3 (coupon side view). Steficek video
101	KB	12/19/01	Gilsanz	K-1	n/a	7/16"	n/a	1/4"	1/4"	3/8"		Chosen due to floor location. Piece marked "93-96".

SEAoNY Summary of Identified WTC Steel Pieces at Salvage Yards as of March 15, 2002

	Yard	Trip Date	Note Taker	Piece ID Mark	d	t_w	b_f	t_{ff}	t_{bf}	t_s	~L	Remarks (description, photos/videos, coupons)
102	KB	12/19/01	Gilsanz	K-2	13-3/4"	0.92"	30"	0.92"	n/a	n/a		Chosen due to fire damage, floor location.
103	KB	12/20/01	Chuliver	K10	12"	9/16"	13"	1/2"	1/4"	1/2"		Chosen due to fire damage. Photo 1-byABonilla. Coupon taken and brought to GMS. Coupon photo by AMassa.
104	KB	12/20/01	Chuliver	K11	14"	1-1/8"	11-5/8"	1/2"	n/a	1/2"		Chosen due to fire damage.Photo 2-byABonilla. Coupon taken and brought to GMS. Coupon photo by AMassa.
105	KB	12/20/01	Chuliver	K12	13-3/4"	7/16"	13"	5/8"	n/a	7/16"		Chosen due to floor location. Photo 3-byABonilla. Coupon taken and brought to GMS. Coupon photo by AMassa.
106	KB	12/20/01	Chuliver	K13	13-11/16"	1/4"	13-1/2"	1/4"	9/16"	3/8"		Chosen due to floor location and fire damage. Photo 5-byABonilla.Coupon taken and brought to GMS. Coupon photo by AMassa.
107	KB	12/20/01	Chuliver	K14	13-1/2"	1/4"	13-3/16"	3/8"	1/2"	3/8"		Chosen due to floor location and fire damage. (Photo:DKoutsoubis:12/27/02:#32-41), Photos 6-7-byABonilla. Stencil mark: floors 91-94. Coupon taken and brought to GMS. Coupon photo by AMassa.
108	KB	12/27/01	Chuliver	K15a	13-1/4"	1/4"	13-1/2"	5/16"	?	3/8"		Chosen due to floor location. Photo 4-by DSharp. Coupon taken and brought to GMS. Coupon photo by AMassa.
	KB	12/27/01	Chuliver	K15b	7-3/8"	3/8"	12-1/4"	3/8"	1/4"			Chosen due to floor location. Photo 5-by DSharp Rectangular piece attached to K15a.
109	KB	1/19/02	Koutsoubis	K16	13-7/8"	1-3/8" to 1-3/4"	11"	1/2"	3/8"	1-1/4"-1-5/8"		Chosen due to unique Thickness variations, due to corrosion or fire? Video Sharp. Photo:DKoutsoubis:1/19/02:#1-13.
110	KB	1/19/02	Koutsoubis	K17	13-5/8"	1-1/2"	10-7/8"	1/2"	5/8"	13/16"		Chosen due to fire damage. Video by DSharp. Photos by Dkoutsoubis 1/16/02:#13-16 and 1/19/02:#14-24.
111	KB	12/27/01	Chuliver	K18	11-1/4"	3/8"	13-1/4"	1/4"	3/8"	1/4"		Chosen due to floor location. Photos by DKoutsoubis:12/27/02:#14-22 and video:12/27/01:2m6s. Coupon taken and brought to GMS. Coupon photo by AMassa.
112	KB	12/27/01	Chuliver	K19	14"	7/8"	12-1/4"	5/16"	n/a	3/4"		Chosen due to fire damage. Photos by DKoutsoubis:12/27/01:#23-27, 30-31 and video:12/27/01:1m24s.Coupon taken and brought to GMS. Coupon photo by AMassa.
113	KB	12/27/01	Koutsoubis	K20	13-1/2"	1/4+"	13	1/4"	1/4"	3/8+"		Chosen due to floor location. Photos by DKoutsoubis:12/19/01:#41-42 and 12/27/01:#1-11, and video:2m3s.
114	KB	1/18/02	Sharp	K40	11"	1/4"	13"	1/4"	1/4"	3/8"		Chosen due to floor location. Photos 59-60 (Hoy)? Marking on piece is 92-95. Also marked 252.

SEAoNY Summary of Identified WTC Steel Pieces at Salvage Yards as of March 15, 2002

	Yard	Trip Date	Note Taker	Piece ID Mark	d	t_w	b_f	t_{tf}	t_{bf}	t_s	~L	Remarks (description, photos/videos, coupons)
115	KB	1/18/02	Sharp	K41	13-3/8"	3/8"	13-1/2"	1/4"	1/4"	3/8"		Chosen due to floor location. Photos 57-58 (Hoy). Marking on piece is "L9901 120".
116	KB	1/19/02	Sharp	K50			See Remarks					Sample of washer failure. Piece photos by DKoutsoubis:1/30/02:#1-5. Piece cut into 3 sections and brought to GMS offices.
117	NW	11/16/01 to 1/18/02	DePaola/ NIST	NIST N-1	13-1/2"	1/2"	13"	3/8"	1/2"	1/2"	3' 8"	Perimeter column. Photo N-1 by NIST.
118	NW	11/16/01 to 1/18/02	DePaola/ NIST	NIST N-2			See Remarks					Piece is comprised of the majority of the pieces which make up a floor truss section from WTC 1 or 2. Dimensions not applicable. Photo N-2 by NIST.
119	NW	11/16/01 to 1/18/02	DePaola/ NIST	NIST N-3	13-1/2"	5/16"	9-1/2"	1/2"			87"	Perimeter column. Photo N-3 by NIST.
120	NW	11/16/01 to 1/18/02	DePaola/ NIST	NIST N-4	13"	3/8"	13"	3/8"	3/8"	9/16"	16'	Perimeter column. Photo N-4 by NIST.
121	NW	11/16/01 to 1/18/02	DePaola/ NIST	NIST N-5	1-3/8"		See Remarks					4' 4" X 11" section of a bolted connection area (spandrel splice) from a perimeter column. Photo N-5 by NIST.
122	NW	11/16/01 to 1/18/02	DePaola/ NIST	NIST N-6	13-1/4"	3/8"	14"	1/2"	1/2"	9/16"	4' 6"	Perimeter column. Photo N-6 by NIST.
123	NW	2/8/02	Koutsoubis	M2	13-1/4"	5/16"	13-1/4"	1/4"	3/8"	3/8"		(Photo:DKoutsoubis:2/8/02:13,15-21) Evidence of burning.
124	NW	2/8/02	Koutsoubis	M3	13-3/4"	5/16"	13-1/2"	1/4"	1/4"	5/16"		(Photo:DKoutsoubis:2/8/02:23-28)(Photo:DKoutsoubis:2/14/02:17-21) Evidence of burning.
125	NW	2/8/02	Hoy	M4	13-1/4"	1/4"	13-1/4"	1/4"	1/4"	3/8"		(Photo:DKoutsoubis:2/8/02:30-34) "Mushed" end.
126	NW	2/8/02	Koutsoubis	M5	14"	1/4"	13-1/2"(*)	1/4"	1/4"	3/8"		(Photo:DKoutsoubis:2/8/02:39-45) Thin spandrel. "d" dimension is approx.
127	NW	2/8/02	Koutsoubis	M6	13-1/2"	1/4"	13-1/4"	1/4"	1/4"	3/8"		(Photo:DKoutsoubis:2/8/02:59-63) Thin spandrel.
128	NW	2/8/02	Hoy	M7	13-1/2"	5/16"	13-1/8"	1/4"	1/4"	3/8"		(Photo:DKoutsoubis:2/8/02:67-72) Evidence of burning.

SEAoNY Summary of Identified WTC Steel Pieces at Salvage Yards as of March 15, 2002

	Yard	Trip Date	Note Taker	Piece ID Mark	d	t_w	b_f	t_{tf}	t_{bf}	t_s	~L	Remarks (description, photos/videos, coupons)
129	NW	2/8/02	Koutsoubis	M8	13-1/2"	1/4"	13-1/4"	1/4"	1/4"	3/8"		(Photo:DKoutsoubis:2/8/02:73-80) Thin spandrel.
130	NW	2/8/02	Koutsoubis	M9	13-1/2"	1/4"	13-1/2"	1/4"	1/4"	3/8"		(Photo:DKoutsoubis:2/8/0281-84) Beat up. "tw" dimension is approx.
131	NW	2/8/02	Koutsoubis	M10	13-3/8"	5/16"	13-1/2"	1/4"	1/4"	3/8"		(Photo:DKoutsoubis:2/8/02:85-92) Burnt on inside.
132	NW	2/8/02	Koutsoubis	M11a	15-1/2"	3/4"	15-3/4"	1-1/2"				W14 Tree Column- column piece. Photos by DKoutsoubis:2/8/02:93-105.
	NW	2/9/02	Koutsoubis	M11b	23-5/8"	5/16"	7"	5/8"				Attached to M11a - beam piece of W14 tree column. Photos by DKoutsoubis:2/8/02:93-105.
133	NW	2/8/02	Koutsoubis	M12	~30"	~1/2"	10-1/2"	11/16"				Severely burnt & twisted wide-flange girder. Studs welded on the top flange. Photos by DKoutsoubis:2/8/02:106-114.
134	NW	2/25/02	Sharp	M13	13-1/2"	1/4"	13-1/2"	1/4"	1/4"	3/8"		Two perimeter columns folded in half Stamped markings "06 92 95 2430". DSharp Photo.
135	NW	2/25/02	Sharp	M14	13-1/2"	1/4"	13-1/2"	5/16"	5/16"	3/8"		Spandrel markings "??? 5T <63> ?? -99-102" Piece is stamped "99-102 3311 102" DSharp photo.
136	NW	2/25/02	Sharp	M15	13-1/2"	5/16"	13-1/2"	1/4"	1/4"	3/8"		Spandrel markings "PONYA A115-89-92 6T" Stamped markings: "89 92 L2616 115" and "L6973 1? 125 36 55".
137	NW	3/1/02	Sharp	M16	13-1/2	9/32"	13-1/4"	1/4"	1/4"	3/8"		3 perimeter columns from WTC 1 Spandrel plate markings "PONYA A148 99-102 <6?>" Piece stamped "102 99" Two small clips on damper.
138	NW	3/1/02	Sharp	M17	~22"	7/16"	12"	3/4"	1-1/8"	n/a	40'	Unique piece with portion of viscous damper attached, also straps and connections. Cut-outs for mechanical. Built-up flange. Marked "163 (9) <62>".
139	FK	12/6/01	Fahey/Rosa	A	36"	1/2"	10-1/2"	2-1/4"			19'-11"	Beam - Appeared heat affected. Flange coupon (A-1) & web coupon (A-2) marked 7'+/- from end of beam. Photo #'s 141, 142 & 475.
140	FK	12/6/01	Fahey/Rosa	B	36"	13/16"	12"	1-1/4"			4'-10"	Beam - Puncture. Web coupon (B-1) marked. Photo #'s 139, 140, 143 & 476.
141	FK	12/6/01	Fahey/Rosa	C	30-1/4"	1"	10"	2-1/2"			14'-5"	Beam - Appeared heat affected. Web coupon (C-1) marked. Photo #'s 135, 137, 138, 144 & 477.

SEAoNY Summary of Identified WTC Steel Pieces at Salvage Yards as of March 15, 2002

	Yard	Trip Date	Note Taker	Piece ID Mark	d	t_w	b_f	t_{tf}	t_{bf}	t_s	~L	Remarks (description, photos/videos, coupons)
142	FK	12/6/01	Fahey/Rosa	D	21-1/4"	1"	17-7/8"	1-9/16"				Column - Puncture. Photo #'s 482 & 483.
143	FK	12/6/01	Fahey/Rosa	E	12-5/8"	3/4"	12-3/8"	1-1/8"				Column - Appeared heat affected. Web coupon (E-1) marked. Photo #'s 488 & 491.
144	FK	12/6/01	Fahey/Rosa	F	38"	3/4"	7"	1-3/4"				Beam - Appeared heat affected. Web coupon (F-1) marked. Photo #'s 490 & 491.
145	FK	12/19/01	Fahey/McConnell	G	35-1/2"	7/8"	16-3/8"	1-9/16"				Beam. Coupon (G-1) marked on moment splice plate. Photo #'s 125, 128 & 129.
146	FK	12/19/01	Fahey/McConnell	H	21"+/-	1/2"	10"	3/4"				Beam. Web coupon (H-1) marked. Photo #'s 131, 132 & 133.
147	FK	12/19/01	Fahey/McConnell	I	7-5/16"	3/4"	15-11/16"	1-1/16"			18'-9"	Brace - (2) WT's back to back - Appeared heat affected. Coupon (I-1) marked on gusset plate at end of brace. Photo #'s 145, 148, 149 & 150.
148	FK	12/19/01	Fahey/McConnell	J	27"+/-	1/2"	10"+/-	1/2"				Beam. Web coupon (J-1) marked. Photo #'s 155, 156 & 157.
149	FK	12/19/01	Fahey/McConnell	K	24-7/8"	3/4"	7"	1-1/2"				Beam. Web coupon (K-1) marked. Photo # 165.
150	GZ	~11/27/01	Eschenasy	-			See Remarks					Connection coupon recovered from WTC#5, 8th floor, and brought to GMS offices. Web tear-out at bolts. Coupon photo by AMassa. Sketch with dimensions and photo in this WTC BPS report, Fig. 4-22.
151	GZ	~11/27/01	Eschenasy	-			See Remarks					Connection coupon recovered from WTC#5, 7th floor, and brought to GMS offices. Web tear-out at bolts. Coupon photo by AMassa. Sketch with dimensions and photo in this WTC BPS report, Fig. 4-22.
152	GZ	~11/27/01	Eschenasy	-			See Remarks					Connection coupon recovered from WTC#5, 6th floor, and brought to GMS offices. Web tear-out at bolts. Coupon photo by AMassa. Sketch with dimensions and photo in this WTC BPS report, Fig. 4-22.
153	GZ	~11/27/01	Eschenasy	-			See Remarks					Connection coupon recovered from WTC#5, 6th floor, and brought to GMS offices. Web tear-out at bolts. Coupon photo by AMassa. Sketch with dimensions and photo in this WTC BPS report, Fig. 4-22.
154	GZ	~11/27/01	Eschenasy	-			See Remarks					Connection coupon recovered from WTC#5, 8th floor, and brought to GMS offices. Plate tear-out at bolts. Coupon photo by AMassa. Sketch with dimensions and photo in this WTC BPS report, Fig. 4-22.
155	GZ	~11/27/01	Eschenasy	-			See Remarks					Connection coupon recovered from WTC#5, 7th floor, and brought to GMS offices. Connection web-plate or splice-block shear. Coupon photo by AMassa. Sketch with dimensions and photo in this WTC BPS report, Fig. 4-22.

SEAoNY Summary of Identified WTC Steel Pieces at Salvage Yards as of March 15, 2002

Yard	Trip Date	Note Taker	Piece ID Mark	d	t_w	b_f	t_{tf}	t_{bf}	t_s	~L	Remarks (description, photos/videos, coupons)
156	GZ ~11/27/01	Eschenasy	-			See Remarks					Connection coupon recovered from WTC#5, undetermined floor, and brought to GMS offices. Column buckling due to heat exposure. Coupon photo by AMassa. Sketch with dimensions and photo in this WTC BPS report, Fig. 4-22.

E *Aircraft Information*

Boeing 767-200ER

General Specifications

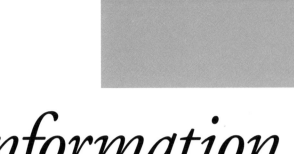

Passengers		
Typical 3-class configuration	181	
Typical 2-class configuration	224	
Typical 1-class configuration	up to 255	
Cargo	2,875 cubic feet (81.4 cubic meters)	
Engines' Maximum Thrust		
Pratt & Whitney PW4062	63,300 pounds (28,713 kilograms)	
General Electric CF6-80C2B7F	62,100 pounds (28,169 kilograms)	
Maximum Fuel Capacity	23,980 U.S. gallons (90,770 liters)	
Maximum Takeoff Weight	395,000 pounds (179,170 kilograms)	
Maximum Range	6,600 nautical miles	
Typical city pairs: New York-Beijing	12,200 kilometers	
Typical Cruise Speed	0.80 Mach	
at 35,000 feet	530 mph (850 km/h)	
Basic Dimensions		
Wing Span	156 feet 1 inch (47.6 meters)	
Overall Length	159 feet 2 inches (48.5 meters)	
Tail Height	52 feet (15.8 meters)	
Interior Cabin Width	15 feet 6 inches (4.7 meters)	

FEDERAL EMERGENCY MANAGEMENT AGENCY

Edward M. DePaola

F *Structural Engineers Emergency Response Plan*

It is recommended that local emergency response plans be developed for natural and manmade disaster events. The plans should define the roles and responsibilities of the parties involved, set up a system of credentialing, and define procedures for the participants. The plan should be flexible and comprehensive to allow for easy modification for differences encountered in each event. The following list, which was taken from the proposed SEAoNY "Structural Engineering Emergency Response Plan (SEERP)," is offered as a basis for the development of other similar plans:

1. Organizations willing to assist in natural or manmade disasters must immediately work with state and local governments to pre-establish a permanent relationship.

2. Issues relating to responsibilities and liabilities of individuals, firms and the organizations must be worked out in advance. These may include a process for the deputizing of volunteers or the enactment of "Good Samaritan" laws.

3. Response teams must be composed of pre-qualified volunteers and not drawn ad hoc from the organization's general membership.

4. There must be specific criteria to ensure that the volunteers are suitably qualified. Each volunteer should also be classified in accordance with their professional knowledge and physical abilities. As a minimum, they must be properly trained in emergency structural evaluations and perhaps urban search and rescue techniques.

5. A database must be created and maintained that contains all volunteer contact and other information.

6. There must be a chain of communication (a tree or pyramid system) for the contact of the volunteers in the event of an emergency.

7. Volunteers must have a specified set of equipment available at all times. Equipment must be tested and maintained regularly.

8. There must be specific, predetermined response locations where the volunteers are to report.

9. There must be a specific photo identification badge which is developed in conjunction with the local authorities. It is recommended that the individual's name, affiliation with the organization and qualifications as a "Structural Engineer" be clearly indicated. It is further recommended that these identifications have specific expiration dates. Periodic renewal will verify compliance with all requirements listed above.

10. The local authority should maintain command and control once the event occurs.

11. The volunteers must be kept actively involved and interested. When the actual disaster strikes, the organizers and the local government need to be assured that the volunteers will be ready, willing and able to help.

12. Volunteers should include experts in other specific disciplines as well. These may include architects, mechanical engineers, civil engineers specializing in underground utilities, and transportation engineers familiar with above- and below-ground transportation.

13. Volunteers for nontechnical assistance such as day-to-day coordination, communications, scheduling, and media relations are also important.

14. A database should be created that contains general information about selected buildings, including data concerning the structural systems, architectural layouts, principal contacts, fire alarm and suppression systems, drawings, and other useful information that would make for a more expedient emergency response.

15. Other Observations:

    - The magnitude of the event may require that plans be modified. Each situation must be evaluated separately. Contingency plans should be in place for all aspects of the response.

    - Situations change rapidly. Plans must be flexible and versatile and participants must have the ability to change with the situation.

    - Good communication equipment on the site is an absolute necessity. Although cell phones may eventually be workable, it must be assumed that normal communications at the event location are not properly functioning. The local authorities must provide high quality two-way radios for teams to keep in contact with each other and the local authorities.

G Acknowledgments

The Federal Emergency Management Agency and the American Society of Civil Engineers would like to acknowledge the contributions made by the following persons to the Building Performance Study of the New York City World Trade Center site.

Report Analysis and Graphics Support

ABS Consulting

Karen Damianick, Project Engineer

Jacques Grandino, Senior Project Engineer

Mark Pierepiekarz, Group Manager

Andre Sidler, Lead Engineer

Daniel Symonds, Senior Project Manager

Arup Fire

James Lord, Fire Strategist

Federal Emergency Management Agency

Arlan Dobson, FEMA Region II, Disaster Assistance Specialist

Gary Sepulvado, FEMA Headquarters, Policy Analyst

Gilsanz Murray Steficek, LLP

Victoria Arbitrio, Associate

Christopher M. Hewitt, Structural Engineer Intern

Joo-Eun Lee, Structural Engineer

Raul Maestre, Structural Engineer

Phillip Murray, Partner

Willa Ng, Structural Engineer Intern

Gary Ray Steficek, Partner

National Institute of Standards and Technology
Ronald Rehm, NIST Fellow, Building and Fire Research Laboratory

Ryan Biggs Associates
Matthew G. Yerkey, Structural Engineer

Severud Associates Consulting Engineers, PC
Andrew Mueller-Lust, Associate Principal
Bill Yun, Associate

Skidmore, Owings & Merrill LLP
Brian McElhatten, Structural Engineer
Juan Paulo Morla, Structural Engineer
Lawrence Novak, Associate Partner
Dean Riviere, Associate
Robert Sinn, Associate Partner
Heiko Sprenger, Structural Engineering Intern

Skilling Ward Magnusson Barkshire, Inc.
Donald Barg, Senior Associate
John Hooper, Principal
Ron Klemencic, Principal

United States Fire Administration
Robert Neale, Training Specialist

Weidlinger Associates
Michael Dallal, Research Engineer
Mohammed Ettouney, Principal
David Ranlet, Senior Scientist

Worcester Polytechnic Institute
Jeremy Bernier, Graduate Student, Mechanical Engineering
Marco Fontecchio, Graduate Student, Mechanical Engineering
Jay Ierardi, Doctoral Candidate, Fire Protection Engineering
Patrick Spencer, Undergraduate Student, Mechanical Engineering/Fire Protection Engineering

APPENDIX G: *Acknowledgments*

WTC Event Data and Summaries

Consolidated Edison
David Davidowitz, Vice President
Stuart C. Hanebuth, Senior Environmental Specialist
Daniel Simon, Manager
Gerard Toto, Field Planner

FEMA Search and Rescue Team
David Hammond, Lead Structural Engineer

Fire Department of New York (FDNY)
Robert Brugger, Deputy Commissioner
Michael Butler, Chief
Sam Melisi, Firefighter
Daniel A. Nigro, Chief of Department

New York City Department of Design and Construction (DDC)
Michael Burton, Executive Deputy Commissioner
Ken Holden, Commissioner

New York City Office of Emergency Management (OEM)
MaryAnn Marrocolo, Director of Recovery and Mitigation
Liam O'Keefe, Emergency Preparedness Specialist
Richard Rotanz, Deputy Director

The Port Authority of New York and New Jersey
Ralph D'Apuzo, Principal Engineer
Joseph Englot, Chief Structural Engineer
Frank Lombardi, Chief Engineer
Dharam Pal, Chief Mechanical Engineer
Jack Spencer, Deputy Chief Engineer

Salomon Smith Barney
James Carney, Vice President
John V. Glass, First Vice President

Verizon
Glen Moyer, Specialist, Design and Construction
Dominic Veltri, Manager, Design and Construction

FEDERAL EMERGENCY MANAGEMENT AGENCY

APPENDIX G: *Acknowledgments*

Photographic and Video Support

American Society of Media Photographers, Inc. (ASMP)
Victor S. Perlman, Managing Director and General Counsel

Peter Skinner, Communications Director

Dick Weisgrau, Executive Director

Federal Emergency Management Agency
Bettina Hutchings, FEMA Region IV, Disaster Assistance Specialist

Here is New York Gallery Photograph Collection
Paul Constantine

Val Junker

Susan Luciano

Jay Manis

Michael Shulan

Charles Traub

Mobius Communications, Inc.
Val Junker, President, Technical Consultant

WCMH
Shannon Harris, Executive Producer

WNBC
Burton Kravitz, Graphic Artist

Dennis Swanson, Vice President/General Manager

Steel Surveying in Salvage Yards

Salvage Yards

Blanford & Co., Keasbey

Lisa Lickman

Ron Lickman, Jr.

Ron Lickman, Sr.

John Sandy

Fresh Kills Landfill

Mark Kucera, USACE

David Leach, USACE

Hugo Neu Schnitzer East
Robert Kelman, General Manager
Frank Manzo, Transportation Manager
Danny Nunes, Heavy Melting Steel Manager
L. Steven Shinn, Operations Manager

Metal Management Northeast, Inc.
Alan Ratner, President
Michael Henderson, General Manager
John Silva, Terminal Manager

Structural Engineers Association of New York – Salvage Yard Volunteers
Amit Bandyopadhyay
Anamaria Bonilla
Peter Chipchase
Anthony W. Chuliver
Edward DePaola
Louis Errichiello
James H. Fahey
Ramon Gilsanz
Jeffrey Hartman
David Hoy
Dean Koutsoubis
Andrew McConnell
Rajani Nair
Alan Rosa
David Sharp
Gary Steficek
Kevin Terry

Report Reviewers

American Institute of Steel Construction
Farid Alfawakhiri
Charles J. Carter
Christopher M. Hewitt
Harry Martin
Thomas Schlafly
Robert Wills

APPENDIX G: *Acknowledgments*

ASCE Technical Council of Forensic Engineering

Paul A. Bosela
Cleveland State University

Merle E. Brander
Brander Construction Technology, Inc.

Leonard M. Joseph
Thornton-Tomasetti Engineers

Robin Shepherd
Earthquake Damage Analysis Co.

Ruben M. Zallen
Zallen Engineering

Council of American Structural Engineers

Joseph C. Gehlen
Kramer Gehlen & Associates, Inc.

Ronald J. LaMere
BKBM Engineers

Raymond F. Messer
Walter P. Moore & Associates, Inc.

Antranig (Andy) M. Ouzoonian
Weidlinger Associates, Inc.

National Council of Structural Engineer Associations

August Domel
Engineering Systems, Inc.

Kurt Gustafson
Tylk Gustafson Reckers Wilson Andrews, LLC

Socrates Ioannides
Structural Affiliates International, Inc.

John Ruddy
Structural Affiliates International, Inc.

Michael J. Tylk
Tylk Gustafson Reckers Wilson Andrews, LLC

National Fire Protection Association

Guy Colonna
Gary Keith
Bonnie Manley
Robert Solomon
Gary Tokle

APPENDIX G: *Acknowledgments*

National Science Foundation Reviewers

Abolhassan Astaneh-Asl
University of California, Berkeley

Lawrence C. Bank
University of Wisconsin-Madison

J. David Frost
Georgia Institute of Technology

Theodore Krauthammer
Penn State University

Antoine (Tony) E. Naaman
University of Michigan

Chief Edward Stinnette
Fairfax County Fire Department

Society of Fire Protection Engineers

Morgan Hurley

Structural Engineering Institute, Technical Activities Division

Reidar Bjorhovde
The Bjorhovde Group

Delbert F. Boring
American Iron and Steel Institute

Theodore V. Galambos
University of Minnesota

Lawrence G. Griffis
Walter P. Moore & Associates, Inc.

John M. Hanson
Wiss, Janney, Elstner Associates, Inc.

William McGuire
Cornell University

Robert J. McNamara
McNamara/Salvia, Inc.

R. Shankar Nair
Teng Associates, Inc.

David B. Peraza
LZA Technology

Robert T. Ratay
Consulting Engineer

APPENDIX G: *Acknowledgments*

Other Selected Reviewers

David Cooper
Flack and Kurtz Inc.

Daniel A. Cuoco
LZA Technology

David Davidowitz
Con Edison

William Grosshandler
National Institute of Standards and Technology

John Healey
Greenhorne & O'Mara, Inc.

Francis J. Lombardi
Port Authority of New York and New Jersey

John Odermatt
New York City Office of Emergency Management

John Sonny Scarff
Marriott Corporation

H Acronyms and Abbreviations

Acronyms

ACARS	Aircraft Communications Addressing and Reporting System
ACI	American Concrete Institute
AISC	American Institute of Steel Construction
APW	air-pressurized water
ASCE	American Society of Civil Engineers
ASMP	American Society of Media Photographers
ASTM	American Society for Testing and Materials
ATC	Applied Technology Council
BBC	British Broadcasting Corporation
BPM	built-up plate member
BPS	Building Performance Study
BRE	Building Research Establishment
BSI	British Standards Institute
CFD	Computational Fluid Dynamics
CIA	Central Intelligence Agency
CM	Claremont (salvage yard)
Con Ed	Consolidated Edison
CTBUH	Council on Tall Buildings in the Urban Habitat
DDC	Department of Design and Construction (NYC)
DEC	Department of Environmental Conservation
DoB	Department of Buildings (NYC)
DOD	Department of Defense
DOE	Department of Energy
EDT	Eastern Daylight Time
EDX	energy dispersive x-ray analysis

FEDERAL EMERGENCY MANAGEMENT AGENCY

EEOC	Equal Employment Opportunity Commission
EPA	Environmental Protection Agency
FCC	Fire Command Center
FDNY	Fire Department of New York
FDS1	Fire Dynamics Simulator Version 1
FEM	finite element modeling
FEMA	Federal Emergency Management Agency
FK	Fresh Kills (salvage yard)
HSLA	high-strength low-alloy
HVAC	heating, ventilation, and air-conditioning
IRS	Internal Revenue Service
KB	Keasbey (salvage yard)
LDEO	Lamont-Doherty Earth Observatory (Columbia University)
LZA	LZA Technology/Thornton-Tomasetti
MTA	Metropolitan Transit Authority
NAIC	National Association of Insurance Commissioners
NBC	National Broadcasting Company
NBS	National Bureau of Standards
NCSEA	National Council of Structural Engineers Associations
NFPA	National Fire Protection Association
NIST	National Institute of Standards and Technology
NJ	New Jersey
NOAA	National Oceanic and Atmospheric Administration
NW	Newark (salvage yard)
NY	New York
NYBFU	NY Board of Fire Underwriters
NYC	New York City
NYPD	New York Police Department
OCC	Operations Control Center
OEM	Mayor's Office of Emergency Management (NYC)
PANYNJ	Port Authority of New York and New Jersey
PATH	Port Authority Trans-Hudson
SCBA	self-contained breathing apparatus
SEAoNY	Structural Engineers Association of New York
SEERP	Structural Engineering Emergency Response Plan
SEI	Structural Engineering Institute
SEI/ASCE	Structural Engineering Institute of the American Society of Civil Engineers
SFPE	Society of Fire Protection Engineers
SOM	Skidmore, Owings & Merrill
SSB	Salomon Smith Barney
STD	Standard
TMS	The Masonry Society
UL	Underwriters Laboratories

APPENDIX H: *Acronyms and Abbreviations*

UPS	uninterruptible power supply
USC	United States Code
UTC	Coordinated Universal Time
WFC	World Financial Center
WTC	World Trade Center
Y/T	yield-to-tensile-strength ratio

Abbreviations

°C	degrees Celsius
°F	degrees Fahrenheit
cfm	cubic feet per minute
ft^2	square feet
gpm	gallons per minute
GW	gigawatt
Hz	hertz
in-k or kip-in	kip-inch
kips/bolt	kips per bolt
kg/m^2	kilograms per square meter
kJ/g	kilojoules per gram
km	kilometer
km/h	kilometers per hour
ksi	kips per square inch
kV	kilovolt
kVA	kilovoltampere
kW	kilowatt
kW/m^2	kilowatt per square meter
m^3/s	cubic meters per second
μm	micrometer
min	minute
mm^2	square millimeter
MPa	megapascals
mph	miles per hour
mW	megawatt
n/a	not applicable
nm	nanometer
nm/s	nanometers per second
psf	pounds per square foot
psi	pounds per square inch
s	second
sf	square foot
V	volt